STEM CENTURY

It Takes a Village to Raise a 21st-Century Graduate

THE SUFFOLK, VA EDITION
SUFFOLK (VA) PUBLIC SCHOOLS

CISTEMIC
PUBLISHING

About Suffolk (VA) Public Schools

Suffolk Public Schools serves over 14,000 students who strive for excellence in education, celebrate diversity, and are committed to students, staff, and the school community.

Located in western tidewater in Hampton Roads, the mission of the school division is

- Produce 21st-century learners that will become productive citizens in society.
- Foster a dynamic, safe, and nurturing learning environment.
- Partner with the school community for the benefit of students and staff.
- Strengthen the school division by employing highly qualified and diverse staff.
- Effectively and efficiently manage capital and human resources.
- Effectively communicate to increase community investment.

Website: spsk12.net
#SPSCreatesAchievers

Contents

Section III - The Practitioners - Workforce Identification and Specification

Section IV - The Practitioners: Central Intelligence

Section V - The Future

Section VI - The School Community

Section 1 - Visionary Leadership

What If I were to tell you that our current instructional curriculums are not helping our students get the jobs that pay the most money?

What if I told you that it IS possible to have STEM incorporated into every subject matter and that in doing so, it creates a cross-curricular approach that has never been done before?

What if it could be read and then seen for all to capture?

ONE

The Vision

Dr. John B. Gordon III

Suffolk Public Schools, located in the tidewater region of Virginia has approximately 95,000 residents and 14,400 students. The student demographics for our school division are 56% African-American, 30% Caucasian, 6% Hispanic, and 6% multi-racial. The school division has 2325 employees, making it the largest employer in the city of Suffolk. The employee demographics are 58% Caucasian and 41% African-American. I have been the Division Superintendent since October 2019, and I had five "regular" months before the COVID-19 pandemic occurred. I can clearly remember making the announcement on March 20, 2020, that the school division was going to "temporarily shut down" due to concerns over staff and student health and safety. When school closures occurred in Virginia, the majority of my colleagues and I thought that it was only going to be for two weeks. Based on conversations that I had with leaders at the Virginia Department of Education and other superintendents, we thought that the two-week closure would allow Governor Northam's office to reassess the pandemic, determine a path forward to keep our schools open and that all we would have to do is adjust our pacing guides and assessment schedules. I was wrong.

The school closures lasted for the remainder of the 2019-2020 school year. In February 2020, we had just begun my first major initiative of *SPS Connect*, which allowed our students to take their Chromebooks home. Suffolk Public Schools had been 1:1 for technology, meaning that each student had a device that they had access to, but their access was only during the school day. Students were not allowed to take their Chromebooks home due to the fear of damage, additional monitoring, and other reasons that did not make a lot of sense to me. I began by allowing our secondary students, those in grades 6-12, to take their Chromebooks home. We felt that our elementary-age students would need additional training on Chromebook care and the typical do's and don'ts for devices. We were in the process of finalizing those procedures when schools were closed. Our secondary students continued their instruction primarily by using *Edgenuity*, a standard-based online learning resource for school districts. Previously, Suffolk Public Schools had used *Edgenuity* for credit recovery; however, we had to quickly think outside of the box, and rework our current user agreement in order to allow our students to learn swiftly how *Edgenuity* worked, and to have our teachers work with the company to ensure that the pacing would allow our students to complete the curriculum.

Because our elementary students were not allowed to take their Chromebooks home, our elementary teachers worked with the teaching and learning team to develop bi-weekly and then tri-weekly instructional packets to reinforce the skills they were taught when school was in session. This lasted for a little over a month until we realized that creating packets with limited new material was not sustainable. As we revised our daily bell schedules to allow for more remediation time, student breaks for social and emotional learning, and teacher planning time, the creation of the packets became an extreme burden on all staff that were involved. We decided to partner with *Scholastic* since the company had already created activities and resources for each elementary subject area and grade level. *Scholastic* was committed to inspiring literacy and learning for all students through traditional, virtual, or hybrid instruction. When we transitioned to *Scholastic*, we also included additional activities, games, instructional materials, and resources that students and families could use throughout the summer. We also created a *Summer Bridge* program as a part of our traditional summer school offerings in order to support students transitioning from one grade level to the next. As an example, the *Summer Bridge* program would introduce first graders to activities and exercises to prepare them better for second grade the following year. Our focus definitely was more on the English and math curriculums, but we soon discovered that science, which is more of a hands-on subject, suffered due to virtual learning.

Our science standards of learning assessments confirmed that students needed to be in the classroom in order to fully master the skills necessary to make science come to life. This is more than just labs that were completed at the secondary level; it also included simple experiments and STEM activities that would normally be a part of the science curriculum maps. When our students returned to face-to-face instruction for the 2020-2021 school year, the school division continued with our plan of "accelerated learning" even though we were in the "hybrid" format of having half of the school attend school for face-to-face instruction every other day. Using the hybrid instruction model was necessary in order to have the proper social distancing of six feet between each student. Classrooms were not large enough to accommodate all students in the classroom, and because of recommendations from the Virginia Department of Health, we had to limit any close contact between students and teachers to less than fifteen minutes. The less than fifteen minutes virtually eliminated the ability to create science experiments and labs that could be completed with those time constraints. We knew that many of our students "learn by doing," which has been a mantra of our science department under the leadership of Dr. Katelyn Leitner, but that had to be adjusted for staff and student health.

Science scores across the Commonwealth of Virginia fell based on instructional data for the 2020-2021 school year. Suffolk Public Schools was not immune to this as a reduced science pass rate for that year was 54% as compared to 59% for the state average. In Virginia, in order to be accredited the minimum pass rate for science must be 70 percent. As we entered the 2021-2022 school year, we felt that our scores would improve based on students being in school for face-to-face instruction every day. However, since we were still dealing with aspects of the pandemic such as staff and student absences due to positive tests and the corresponding quarantine that may have been necessary based on individual vaccination status, there were still interruptions in learning. In order to combat some aftereffects of learning loss, Suffolk Public Schools continued with our Saturday Academy which would meet every other Saturday for enrichment activities to improve student mastery of skills. During the 2021-2022 school year, we offered courses such as kindergarten through eighth-grade mathematics, Science in grades 6-8, as well as Algebra I, Biology, and Algebra II. Due to the continued health and safety concerns, we also partnered with *Tutor U,* through the Public Consulting Group, in order to offer these same courses in a virtual format. These STEM-related courses were chosen based on student assessment data and feedback from our teachers and members of the teaching and learning team. During the 2021-2022 school year, we offered a total of ten sessions and had 1287 students that participated. We felt that using the Saturday Academy would be another layer of instructional support for our students as it was open to all of our kids, but some were strongly recommended to attend based on their current student progress in the classroom. Along with other instructional supports that were put into place, the science performance of all students improved to 58% based on student achievement data for the 2021-2022 school year, as compared to an average science pass rate of 65% at the state level. Even though the school division improved, we were still below the required 70% pass rate for science, and the gap between the school division pass rate and the state average increased. We had to do something different.

In this book, you will read about additional instructional strategies and supports that we put in place for science instruction. The chapter by Mrs. Kelly Greening, Coordinator of Mathematics Instruction, will also detail how Suffolk Public Schools had one of the highest gains in math student achievement in the state with a twenty-one percent increase from the 2020-2021 to the 2021-2022 school year. The partnership between Dr. Leitner, Coordinator of Science Instruction, and Mrs. Greening is phenomenal and has laid the foundation for an increase in STEM engagement throughout the school division and school community. But we must do more. If the school division is serious about becoming the premier school division in the country, it will be necessary to be innovative in connecting all of our schools, developing a stronger partnership with our community, and introducing our students to careers at an earlier age in order to fulfill the jobs that are still vacant in the computer science field.

Mr. Skinner, Director of Career and Technical Education, will detail our work with economic and workforce development, and how we are providing industry and workforce credentials for more students. It is also our goal to be able to share our plan to improve science assessments with other school districts and document our progress along the way. Led by Dr. Okema Branch, Chief Academic Officer, you will also read about how we developed cross-curricular maps for all subject areas, with student interest being at the center of enrichment activities. Nowhere is it written that STEM can only occur in math and science. Suffolk Public Schools wanted students to transfer the skills that they developed in a specific subject to all subjects. As a part of our Continuous Learning for Continuous Improvement model, the school division focused more on analyzing the data for specific cohorts of students, having individual staff members lead professional learning in order for teachers to hear more from peers instead of division leaders, and finally having conversations with students to determine what worked.

The vision for *STEM Century: The SPS Edition* will not only provide "the recipe" for improving student achievement in science, but it will also give direction to STEM enrichment, improved career interest in STEM careers, and highlight the "TEAM SUFFOLK" approach to collaboration at all levels of the educational experience. In order for this vision to become a reality, I need everyone who is reading this book to tell someone about our plan for providing STEM professional learning that not only will improve student achievement, but will also increase opportunities for students to obtain internships, and obtain experience in employment opportunities that are right around the corner. The goal is for Suffolk, Virginia and Suffolk Public Schools to lead the way in STEM career development.

As you read this book, you will see small narratives between chapters that will serve as a bridge in your reading and in your thinking so that you can get the full perspective on the vision.

About The Author

DR. JOHN B. GORDON III

Dr. John B. Gordon III is the division superintendent of Suffolk (VA) Public Schools. A lifetime educator with leadership experiences in Chesterfield, Fredericksburg, and Richmond, Dr. Gordon understands how education should be customized to fit the needs of all students while expanding opportunities for their future.

As a former successful high school basketball coach, he is an expert in team building and strategic planning and serves as a role model to today's youth. Dr. Gordon was the first African-American principal at James Monroe High School and the first African-American chairman of the Virginia High School League. A renowned speaker and motivator, Dr. Gordon is the author of *The Teacher's Lounge: The Real Role of Educators in Your Schools*, which navigates life as an African American male in predominantly white suburbia. He also serves as an adjunct professor at the University of Virginia and is president of Schools That Inspire, LLC, where he develops transformational strategies for students, teachers, and instructional leaders. Dr. Gordon was recently named one of the 2022 Superintendents to Watch by the National School Public Relations Association.

Dr. Gordon earned a bachelor of arts in psychology from the University of Virginia, a master's of education with a concentration in social studies from the University of Virginia, a post-master's certificate in educational leadership from Virginia Commonwealth University, and a doctorate in educational leadership and policy studies from Virginia Tech. He has three children, Marcus, Kennedy, and Simone.

Website: https://schoolsthatinspire.com/
Suffolk Public Schools Website: https://www.spsk12.net/

TWO

Workforce Development: School is a Great Place to Start

Marlon Lindsay

THE FUTURE OF JOBS, THE FOURTH INDUSTRIAL REVOLUTION, a report by the World Economic Forum, urged that developments in fields such as artificial intelligence, robotics, nanotechnology, 3-D printing, genetics, and biotechnology, will cause widespread disruption to business models and labor markets over the next five years. Tremendous change is predicted in the skill sets needed to thrive in the new landscape.

We are only 23 years into the 21st-century with 77 more years to go, and we are seeing evidence of disruption all around us. We now have AI doing therapy, practicing law, driving cars, and replacing radiologists. Electric cars are putting gas stations and mechanics out of business. On top of that, we have robots laying bricks, installing sheetrock, and cutting lawns. Houses are being 3D-printed in 24 hours. AI is writing research papers and newspaper articles. In fact, when was the last time you saw a paperboy/girl? And this is just the tip of the iceberg.

Globally, there is an increasing shortage of workers for STEM-related jobs. Countries will be adversely impacted if immediate and lasting measures are not employed to counter the displacement of traditional jobs due to the new technologies. For communities to thrive in a 21st-century digital economy, they must include the alignment of education, skills and talent development, and workforce development. This makes for a strong economic development foundation.

Training a workforce cannot be an afterthought for schools and industry. There must be a systemic approach to solving the workforce shortages beginning in schools. Ironically, as I discussed in the first *STEM Century* book, education was organized for work, but work has changed, and now education must do the same to ensure that our students are ready for what lies ahead. It is imperative that we develop 21st-century skills in students, shift the culture of education for school teachers to prepare a 21st-century graduate, and include the community in ways we have not before as a part of a student's educational journey. From the community, we can liaise with industry partners to provide job shadowing, internships, apprenticeships, and essential on-site work experiences to help students hit the ground running after graduation.

While industry partners provide obvious real-time job experience, often overlooked are the opportunities for workforce development within the school system itself. One of the most important benefits of having a job working within the school system is that it can help students develop 21st-century skills. These are the skills essential for success in a 21st-century economy, and include things like critical thinking, problem-solving, communication, and collaboration. Students having jobs while still in school gives them a chance to practice these skills in a real-world setting, which can be very valuable. Working hard at a job helps students develop a strong work ethic while learning the importance of punctuality, responsibility, and other key workplace skills, such as working with a supervisor and peers.

Working with others can teach students how to communicate effectively, collaborate, problem-solve, and build relationships while being adaptable and resourceful. These are all skills that will be valuable in any industry and in any career. There is a wide range of jobs available for students in school, and each one can offer different benefits.

Schools are increasingly replacing contractors and outside workers with students. This trend is evident in everything from food service to maintenance and janitorial work. The main benefit of this shift is the cost savings. With so many schools struggling to make ends meet, any and all savings are welcomed. With the need for students to develop 21st-century skills and workplace experience in real-time and the need for schools to save on expenses, it is a no-brainer to develop a student workforce.

While we can see students replacing contractors and outside workers, we might have issues with students teaching. In the United States, and around the globe there is a tremendous teacher shortage. Many people believe that the answer to this problem is to pay teachers more money, and I agree! However, there are additional solutions to help remedy this problem. One way is to use students to solve the teacher shortage. First, it is a great way to get more people interested in teaching. Second, it gives students real-world experience. And of course third, it is a way to save money. While these are just some of the basic benefits, there are some risks to using students to solve the teacher shortage. First, students may not be as effective as teachers. Second, students may not be able to handle the workload. Third, students may lack the maturity and interest to teach, leading to a lot of turnover.

It is not lost on me that using students to teach is controversial and for many, is the sacred cow. Despite the risks, using students to solve the teacher shortage is a good idea. However, let's look at examples of jobs that develop 21st-century skills in each category.

Communication and collaboration: In the vein of students as teachers, a peer tutor is one job that incorporates communication and collaboration skills. As a peer tutor, a student would be responsible for helping other students in their class who may be struggling with the material. This job would require students to communicate effectively in order to explain the material in a way that is easy for the other students to understand. In addition, students would need to collaborate with the other students to come up with different ways of approaching the material.

Critical thinking and problem solving: One job that requires critical thinking and problem-solving skills is a research assistant. As a research assistant, students would be responsible for helping a teacher with a research project. This would involve students going through different sources of information and critically evaluating them in order to extract the most relevant information for the project. Students would need to be able to solve problems that arise during the research process.

Creativity and innovation: A job that allows students to be creative and innovative is being a graphic designer. As graphic designers, students would be responsible for creating visuals for the school website, yearbook, or other school publications. This would require students to come up with creative ideas and then execute them in a way that is visually appealing.

Digital literacy: Being a social media manager requires digital literacy skills. As a social media manager, students would be responsible for creating and managing the school's social media accounts. This would involve students knowing how to use different social media platforms, as well as how to create engaging content. In addition, students would need to be able to monitor the school's social media accounts and respond to any comments or questions.

Social and emotional learning: A job that incorporates social and emotional learning is being a mentor. As a mentor, students would be responsible for helping younger students adjust to school life. This would involve students being able to listen to the younger student's concerns and provide guidance and support. In addition, students would need to be able to model positive social and emotional behavior for the younger students.

Beyond the aforementioned highlights, the following are jobs that students, when properly trained, are more than capable of performing with the proficiency that closely match the professional:

1. Social Media Manager: The school can appoint a student to manage all the social media accounts. The duties of a social media manager would be to post updates about the school, create engaging content, monitor feedback and respond to queries. This is a great opportunity for a student to develop their marketing and communication skills.

2. Event Planner: From organizing school fairs to fundraisers, a student event planner would be

responsible for coming up with ideas, coordinating
with different stakeholders and making sure that
everything runs smoothly on the day of the event.
This job would suit a student who is creative,
organized, and strong at multitasking.

3. Office Assistant: The office assistant is responsible
   for various administrative tasks such as answering
   phone calls, filing documents, photocopying, and
   scanning. This job is ideal for a student who is detail-
   oriented and with good communication skills.

4. Graphic Designer: The school can appoint a student
   graphic designer to create visuals for the website,
   social media, marketing materials, and more. This is
   a great opportunity for a student to showcase their
   creativity and develop their design skills.

5. Website Manager: The website manager would be
   responsible for keeping the school website up-to-
   date, adding new content, ensuring all links work,
   and responding to queries. This job would be
   suitable for a student who is web-savvy and has good
   writing skills.

6. IT Support Technician: The IT support technician
   would be responsible for providing technical
   assistance to staff and students. This job would
   involve troubleshooting issues, installing software,
   setting up devices, and more. This job would be
   perfect for a student who is interested in technology
   and has good problem-solving skills.

7. Data Entry Clerk: The data entry clerk would be
   responsible for inputting data into the school's
   system. This would involve tasks such as transcribing
   marks, updating student records, and more. This job
   would be suitable for a student who is detail-oriented
   and has good computer skills.

8. Customer Service Representative: The customer service representative would be the first point of contact for parents and students. This would involve answering phone calls, responding to emails, handling customer inquiries, and more. This job would be perfect for a student who has good people skills and is able to handle customer complaints in a professional manner.

9. Accounts Receivable Clerk: The accounts receivable clerk would be responsible for managing the school's accounts receivable. This would involve tasks such as invoicing, issuing refunds, following up on payments, and more. This job would be suitable for a student who is detail-oriented and has good math skills.

10. Fundraiser: The fundraiser would be responsible for coming up with ideas to raise money for the school, coordinating with different stakeholders, executing fundraising events, and more. This job would suit a student who is creative, organized, and good at multitasking.

11. Technology Assistant: With more and more schools using technology in the classroom, there is a growing need for students who are tech-savvy that can help with technical issues. As a technology assistant, the job would be to help teachers and students with anything from using the school's learning management system to setting up audio-visual equipment for presentations.

12. Marketing and Communications: Most schools have a website and some form of print communication, like a newsletter or yearbook. As a marketing and communications assistant, your job would be to help with these projects. This could involve writing articles, taking photos, designing layouts, and more.

13. <u>Recycling and Sustainability Assistant</u>: Many schools are now focused on recycling and sustainability initiatives. As a recycling and sustainability assistant, your job would be to help with these programs. This could involve sorting the recycling, composting, conducting waste audits, and more.

14. <u>Peer Mentor</u>: Many schools have peer mentor programs which pair older students with younger students. As a peer mentor, your job would be to be a friend and resource for your mentee. This could involve meeting regularly, helping with homework, and just being someone to talk to.

15. <u>Data Collector</u>: With all the data that schools collect, there is always a need for students who can help input and organize it. As a data collector, your job would be to input data into the school's database, whether it's student information, test scores, or attendance records. This could be a perfect job for someone who is detail-oriented and good with computers.

16. <u>Tutor</u>: Many students need help with their studies, whether they're struggling with a certain subject or just need extra help to boost their grades. As a tutor, your job would be to provide one-on-one or small-group instruction to students who need it. This could involve working with students before or after school, during lunch, or even during study hall.

And because I really want to drive the point home and give you actionable ideas, here is a more exhaustive list of jobs for K-12 students to develop 21st-century skills:

1. Start a recycling program at your school

2. Work with the school's janitorial staff to develop more efficient cleaning methods
3. Conduct energy audits of classrooms and other school buildings
4. Implement a composting program for the school cafeteria
5. Help plan and execute school-wide events such as field days or pep rallies
6. Serve as a tour guide for prospective students and their families
7. Create a social media marketing campaign for the school
8. Develop a marketing plan for the school's sports teams
9. Create and maintain the school's website
10. Write articles for the school newspaper or yearbook
11. Serve as a photographer for the school newspaper or yearbook
12. Serve as a videographer for the school's sports teams
13. Create a promotional video for the school
14. Write and produce a school podcast
15. DJ school dances or other events
16. Create and sell spirit wear for the school
17. Plan and execute a fundraiser for the school
18. Help with the school's after-school program
19. Serve as a tutor for other students
20. Mentor a younger student
21. Help with the school's garden
22. Work in the school cafeteria
23. Help with event set-up and clean-up
24. Serve as a crossing guard
25. Help with the school's recycling program
26. Work in the school library
27. Serve as an office assistant

28. Help with the school's landscaping
29. Shovel snow in the winter
30. Paint classrooms or other school buildings
31. Help with the school's groundskeeping
32. Perform simple maintenance tasks around the school
33. Serve as a messenger between different parts of the school
34. Help organize and file paperwork
35. Serve as a hall monitor
36. Help with the school's emergency preparedness plan
37. Serve as a safety patrol
38. Help with the school's lost and found
39. Serve as a liaison between school initiatives and community/family
40. Serve as a liaison for connecting with local businesses and the school

In conclusion, using student labor for school projects can have several benefits, both for the students, school systems, and for workforce development. Here are some of the benefits of using student labor for school projects:

1. **Skill Development:** When students work on school projects, they can develop a range of skills that will be useful for their future careers. For example, they may learn how to work collaboratively, how to manage time effectively, and how to communicate with others. These skills can help them succeed in their future work roles.

2. **Real-World Experience:** Working on school projects can provide students with real-world experience, helping them understand how the concepts they are learning in the classroom apply to real-life situations. This can make their education

more relevant and engaging, as they see the practical applications of what they are learning.

3. **Portfolio Building:** Students can use the work they do on school projects to build a portfolio of work that they can showcase to potential employers. This can help them stand out when they are applying for jobs, as they have evidence of their skills and accomplishments.

4. **Workforce Development:** Using student labor for school projects can have a positive impact on workforce development. By giving students the opportunity to work on projects that are relevant to their future careers, schools can help prepare them for the workforce. This can create a pipeline of talent for employers, who may be more likely to hire individuals who have already demonstrated their skills and knowledge in a real-world setting.

5. **Cost Savings:** This one is tricky and sensitive but, I am assuming we are not exploiting our children and that the proper nurture, caring, experience, and pay are provided. One of the most significant benefits of using student labor for school projects is that it can be cost-effective. As apprentices and interns, students aren't qualified to receive pay at the same level as experienced professionals, making it a more affordable option for schools with limited budgets to save on cost while providing students with authentic work experience.

Our young people are brilliant. With the proper preparation, guidance, and opportunities, we can prepare a 21st-century graduate ready to take on our world's greatest challenges in the rest of the 21st century and beyond. We must fully prepare them for the workforce, college, entrepreneurship, and service. We have them for 13 years before they are released into the world, there are no reasons for them not to be ready, and school is a great place to start. Let's get them ready!

~

References:

Closing The Skills Gap: Creating Workforce-Development Programs That Work For Everyone.

Martha Laboissiere and Mona Mourshed, McKinsey & Company, 2017

6 Facts About America's STEM Workforce And Those Training For It.

By Brian Kennedy, Richard Fry and Carey Funk, PEW Research 2021

4 Ways To Bridge The Global Skills Gap, HBR

by Mihnea Moldoveanu, Kevin Frey, and Bob Moritz, Harvard Business Review

3 Ways To Disrupt Education And Help Bridge The Skills Gap, WEF, August 2021

B. Kalyan Kumar, Global Chief Technology Officer and Head, Ecosystems, HCL Technologies

Rhode Island, North Dakota Work to Bridge Work-force Skills Gap By Casey Leins, USNews

About the Author

MARLON LINDSAY

Marlon is a Jamaican-born business leader and author who is committed to helping people reach their full potential. He is the Founder and CEO of **21stCentEd** which provides Comprehensive STEM™ education to school districts and their communities across the United States. He is also a best-selling author of *STEM Century: It Takes a Village to Raise a 21ˢᵗ-Century Graduate* in collaboration with over twenty school superintendents across the United States, and sits on the Utah Valley School of Education Advisory Board.

Since the COVID-19 pandemic, he has doubled down on ensuring that *all* students, especially the underserved and underrepresented, have a foundation in STEM by bringing affordable and scalable comprehensive STEM education to school districts and their communities.

After immigrating to the United States from Jamaica at the formative age of thirteen, Marlon grew up in Connecticut, where he earned his undergraduate and master's degrees from the University of Connecticut. He furthered his education in leadership, financial management, and business development, at Wharton, Columbia, and Dartmouth respectively. He sits on several boards and is excited about the work being done as a trustee of Mountainland Technical College

(MTech) in Utah and the Jamaica STEM Foundation in Jamaica.

Marlon lives in Utah with his wife, Amanda, where they are the proud parents of Skye, Makenna, Bliss, Shiner, River, and Wilder.

Section II - The Practitioners: Economic Exploration

It is the goal of all school divisions to be able to produce well-rounded students that will become 21st-century citizens armed with the tools necessary to become productive citizens in society with the skills necessary to develop and support the workforce. In Suffolk Public Schools, it is our vision to strive for excellence in education, celebrate diversity, and be committed to students, staff, and the school community.

This commitment includes our elementary students exploring their future careers and developing an interest in school and the subject matters they love. At the middle school level, the staff supports student selection and identification as they begin to take a deeper dive into their core subjects while learning that in our STEM model, subjects will no longer be taught in isolation. Finally at the high school level, students begin to specialize based on their interests and exposure to career opportunities. With the support from the School Administration Office, students and staff receive resources that foster learning while building stronger connections to the community.

ONE

The Problem of Practice

Dr. Okema Branch

Science, Technology, Engineering, and Mathematics (STEM) education is lively and a part of our daily existence. Each STEM area is its own force that drives every aspect of our lives, from cell phone usage, weather predictions, telling time, budgets and finance, biological functions, transportation, and most of our lives. STEM impacts everything we do and cannot be separated or compartmentalized to select when we want to use and learn it.

However, if STEM is such an important part of our everyday lives and in everything we do, why limit the understanding and learning of it in school? We want students to engage in their learning and increase deeper learning by developing the Profile of a Virginia Graduate's Five C's - critical thinking, creative thinking, collaboration, communication, and citizenship. So why are classroom structures and pedagogical practices so traditional and limited to compartmentalized content? According to the 2022 Virginia Association of Supervision and Curriculum Development (VASCD) Profile of a Virginia Classroom publication, deeper learning not only equips students with the understanding and skills needed to solve problems and explore questions they have not encountered before but also leads them to metacognition - awareness and understanding of how they think and learn. How educators design authentic experiences, provide opportunities for inquiry, and tailor learning experiences impact the knowledge and skills that support students' readiness for life beyond school. What happens in the classroom with instruction and interactions can greatly impact the next phases of learning and life for students.

Suffolk Public Schools traditionally rotated elementary science and social studies instruction bi-weekly. Teachers were given a pacing guide that led them through each content area and they were to plan accordingly. While several of the science and social studies units took longer than the two-week timeframe, teachers were following pacing and adhering to what had been directed. The results of this are students not retaining the information and teachers having difficulty incorporating the investigations. The current elementary instructional schedule integrates science and social studies alternatively within the 130 minutes per day of English/Language Arts (ELA) and 70 minutes per day of mathematics. Due to the dedicated time allotted for ELA and mathematics, in addition to a cultural mindset of limited time to get everything done, this decreased or even eliminated labs and hands-on learning that would have aided students in using the scientific process and applying the learning and discovery. At the time, it also was not required for teachers to facilitate science labs and investigative activities. At the secondary level, labs were recommended but not required or monitored regularly. Ostroff (2016) writes, "The prevailing cultural myth of our era is that we are forever short on time.

And that feeds into corollary myths: that educators must prioritize content rather than build understanding in order to prepare students for success on examinations and tests; that quantifiable achievement trumps slowing down and genuinely engaging." STEM instruction is as critical and needs the application of cross-curricular instructional consideration as all other contents.

Additionally, teachers received professional science learning during designated times of the year and were limited in their own exploration of the science topics they teach. Pre-service, city-wide Professional Learning Communities (PLSs), and a professional learning day were the only real opportunities for teachers to dive into science education and the pedagogical approach to instruction, and only for a limited time. A division culture of content-specific focus and assessments, without a heavy emphasis on instructional practices that lead to student achievement, have created a conundrum of habits, mindsets, and practices that are evidenced in low science scores and student outcomes.

With the focus on student learning loss amid a return to the classroom from COVID-19 closures, the emphasis on learning and acceleration was again placed on reading and mathematics, with little to no consideration for integrating science and scientific skills into daily instruction. It was not considered, or not evaluated, how limiting science and social studies learning impacted students' overall learning and achievement. Science and social studies are not separate subjects to be picked up when there is time, but they both play a part in learning every other content and being able to apply information. COVID required school closures, and as a results, science instruction was limited to information-only, with little to no science exploration at home. Staff were limited in the materials they had on hand and could not control what students may or may not have had in their homes to apply and explore their learning. Science materials being sent to homes were also limited. For most of the COVID-19 period of uncertainty in schools and reopening variations, much of what was done in science education was limited to lecturing basic science concepts.

Now that we are back in school and realize the great impact the limited delivery of science instruction has had on student outcomes, the division must address the data and what has been discovered in our system and practices. In the article, "Why Science?" author Leah Shafer writes, "educators should portray science as acquiring skills, rather than memorizing facts. If the classroom focuses on the scientific process of discovery, more students will be engaged in the subject matter. Teaching science should be much more than the rote memorization of theories, formulas and vocabulary. It should be an education in problem solving and collaboration" (Shafer, 2015). How we teach science and integrate it across the curriculum will impact student inquiry, learning, and knowledge. Zaretta Hammond notes that helping students build an academic vocabulary in math and science will lay a strong foundation for doing more rigorous conceptual thinking in those subject areas (2015). Suggesting that how teachers teach science and math and the academic vocabulary in those subjects is integral to a student's ability to process information, and build intellectual capacity and independence. How culturally relevant the concept is to their lives and how a teacher presents the vocabulary and background, not only for memorization, but also application, expands a student's cognitive power and prepares them to be capable of taking on any academic challenge.

George Phillips's book, *Math is Not a Spectator Sport (2005)*, focuses on the active exploration of learning math concepts versus a passive reception of formulas and information. Active student learning can lead to higher achievement. How are we actively engaging students in STEM education? How we teach and explore concepts matters. Teaching has traditionally looked like the age-old "sage on the stage", with the teacher leading and lecturing with hopes and expectations of learner retention of information to produce outputs for assessments. No real-life application or relevance to connect and absorb information, but passively receiving to produce later. Greater emphasis should be on students having to develop, read, solve, create, analyze, and summarize, with less time allocated for teacher presentation and talking. Active student-centered learning is staged for information flowing in both directions as students are discovering and applying information, investigating their new knowledge, and testing their own hypotheses based on prior knowledge and exploration. Active student-centered learning is not passive, with simply receiving information to regurgitate for an assessment, but the constant generation of new and unknown concepts that are culturally relevant, peak interests and inquisitiveness, and force students to engage in their own learning.

In a study noted by Rollins (2017), there was a direct correlation between student engagement and instructional methods. The lowest level of engagement, 54.4 percent, occurred when teachers were talking. Even though the lecture was the least engaging delivery method, it was the dominant one. In contrast, students were the most engaged when they were working in labs (73.3 percent) and in groups (73 percent). So the methods that worked best for learning were used the least and the methods that worked the least were used the most. If educators continue to ignore best practice instructional approaches, our students will continue to show a decline in scientific knowledge, but also in knowledge retention and application of all contents.

Our overarching consideration in building strong, inherent, and sustainable science education practices while developing 21st-century skills in all of our students begs the question hat are we teaching and how are we teaching it to captivate interests and inspire lifelong learning and skills in our classrooms? As a division, we recognize the impact of cultural and systemic practices, time restraints, and varying schedules that limit opportunities for relevancy in concept retention and application as well as hands-on learning. We also acknowledge the work necessary for improvement and will begin the foundational work by prioritizing our teaching and learning focus on the following:

- Continued administrator instructional leadership capacity building
- Continued teacher instructional capacity building
- Increased professional learning opportunities for teachers
- Revisiting the science and social studies scheduling at the elementary level
- Revisiting lab and exploratory activities requirements at all levels.

Suffolk Public Schools is positioned to be a premier school division on the cutting edge of innovative and sustainable teaching and learning practices along with systems and processes that directly influence 21st-century students who are prepared to be global leaders and influencers in every field. Our response to constant continuous learning for continuous improvement will drive our efforts and help us overcome barriers and maximize progress in areas of opportunity. We are poised to go to the next level and beyond.

References:

Hammond, Zaretta. *Culturally Responsive Teaching and the Brain*. Thousand Oaks, Corwin, 2015.

Ostroff, Wendy. *Cultivating Curiosity in K-12 Classrooms*. Alexandria, VA, ASCD, 2016.

Phillips, George. *Math is Not a Spectator Sport*. Springer, 2010.

Shafer, Leah. Why Science? *Usable Knowledge*, 13 November 2015, https://www.gse.harvard.edu. Accessed February 3, 2023.

VASCD (April 3, 2022). *Profile of a Virginia Classroom*, 2nd Ed.

About the Author

DR. OKEMA BRANCH

Dr. Okema Branch is a strategic and passionate educational professional who has a unique value proposition of extensive experience in the development, facilitation, and oversight of academic programs and initiatives that promote student growth and achievement, as well as direct teaching experience in both K-12 and higher education establishments, with additional expertise in educational human resources.

Dr. Branch has senior leadership experience and a track record of success serving large diverse student cohorts, collaborating with critical external agencies and partners, developing engaging and impactful outreach initiatives and generating funding, managing sizable budgets and teams, and coordinating professional development opportunities for both faculty and non-educational staff.

Dr. Branch is highly adept in the development, implementation, evaluation, and continuous improvement of academic programs, policies, procedures, and challenging inspiring curricula. Dr. Branch currently serves as the Chief Academic Officer of Suffolk Public Schools. Her role encompasses all aspects of student instruction and academics, assessment, teacher professional learning, and facilitating leadership development.

TWO

STEM Revisions

Dr. Maria Lawson-Davenport

The Challenge of COVID-19

In the spring of 2020, Suffolk Public Schools (SPS) entered an unknown and challenging time in education. COVID-19 struck and closed schools indefinitely for the 2019-2020 school year. SPS was not immune to the chaos and uncertainty that came from these unforeseen school closures and like many divisions, we moved to provide resources and instruction to students for the remainder of the school year. Like many divisions throughout Virginia, we were unable to assess our students through our annual state-mandated Standards of Learning Exams, and we closed out the school year with hopes that September would see an ease to the virus's hold on the world and the return of students to our classrooms. Of course, that didn't happen, and SPS leaders pivoted to online learning.

While there is much that came out of COVID-19 school closures and online learning, the most significant shift was in *how* teachers taught. We moved teachers behind a screen or shield and limited their interaction with students to ensure safety. Teachers learned how to navigate virtual meet platforms like Google Meet and Zoom, while simultaneously delivering as much content as possible via a learning management system. While technology was our greatest asset during this time, it led to some unforeseen habits in instruction that we recognized as a division that had to be unlearned.

Tales of a Framework

As SPS closed its buildings in 2020, the Teaching and Learning Department was finalizing a framework to guide instruction for the upcoming 2019-2020 school year. A process that was three years in development, we had collaborated across departments to outline the core characteristics of planning, teaching, and engaging students to create students who demonstrated the attributes of a Profile of a Virginia Graduate. The Profile of a Virginia Graduate emphasizes the development of the 5Cs - critical thinking, creativity, communication, citizenship, and collaboration. Our team developed the Teaching and Learning Framework as a model for effective instruction that included the foundation for planning lessons, delivery of lessons, and what students would gain from instruction. It was an all-encompassing framework that included instruction and assessment, culturally responsive practices, social-emotional learning, and positive behavior interventions.

The Teaching and Learning Framework was developed and finalized in February 2020, one month before schools shuttered their doors for the school year. In the midst of school closures and the need to adjust instruction, SPS put the Teaching and Learning Framework on hold to better support teachers where they were at the moment. We produced guidance documents for virtual instruction, designed models for flipped, hybrid, and synchronous classrooms, and set expectations for teacher/student engagement. While the focus was not on the framework, it was an ever-present blueprint that we planned to revisit once things returned to normal. However, we began to recognize that normal did not exist in education anymore.

Out of the Box-Relearning Instruction

In the 2021-2022 school year, the SPS Curriculum and Instruction (C & I) Team worked hard to assist teachers with the return to the classroom from the online, virtual learning of the previous year and a half. As the Director of Curriculum and Instruction, I began to see how the trauma of the quick pivots had taken its toll on the instructional practices of most of the teachers in SPS. According to research by Etchells et al. (2021) and Ozamiz-Etxebarria et al. 2021, educators struggled during the pandemic with mental fatigue which led to "pressure from increased expectation to keep the education system afloat and feeling helpless and detached due to the virtual learning environment." This stress was the unfortunate result of teachers having to adopt new methods of instruction with limited support via material or human resources. While teachers also learned how to adapt and overcome the school shutdown, they also placed themselves in an instructional box that was limited in interaction and direct instruction. Once we returned to the classroom, teachers were not only nervous about a full return to the classroom, they were also confused about how to take the lessons learned from relying on technology and merge them with the "normal" instruction. Our team of instructional experts supported teachers with resources and guides to address the gaps in learning that occurred. We narrowed our pacing to reflect "power skills" needed to bridge the gaps between grade levels. We define power skills as those key foundational skills based on grade level or course expectations. These skills were outlined by the C & I team and the Virginia Department of Education for the areas of math and English. We planned with teachers, coached them through instruction, modeled lessons, and disaggregated data to help with small group instruction. While it didn't seem like enough, it was enough to help our students demonstrate growth in areas like math and reading in various schools across the division at the end of the 2021-2022 school year.

Source URL: https://schoolquality.virginia.gov
1/31/2023, 4:26:04 PM

Reading Performance: All Students

Reading results for 2019-2020 are not available due to the closure of school and cancellation of state assessments. 2020-2021 reading results reflect reduced student participation in state reading assessments due to COVID-19. The wide variations in participation rates and learning conditions should be taken into consideration when reviewing 2020 2021 data.

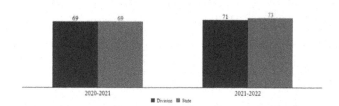

Figures 1.1 Suffolk Public Schools Reading Performance

Source URL: https://schoolquality.virginia.gov
1/31/2023, 4:26:49 PM

Math Performance: All Students

Mathematics results for 2019-2020 are not available due to the closure of schools and cancellation of state assessments. 2020-2021 math results reflect reduced student participation in state math assessments due to COVID-19. The wide variations in participation rates and learning conditions should be taken into consideration when reviewing 2020-2021 data.

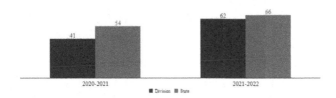

Figure 1.2 Suffolk Public Schools Math Performance

Despite our modest growth as a division when school reopened, the Curriculum and Instruction Team recognized that teachers had forgotten some of the basics of effective tier-I instruction. Tier I instruction is defined as the evidence-based instructional practices that meet the needs of a wide range of learners-at least 80% of a class learns during tier I instruction. As a division, we define tier I instruction by the following characteristics:

1. Standards-aligned research-based curriculum/competency aligned to 5Cs
2. Focus on literacy skills
3. Focus on numeracy skills
4. Evidenced-based & high yield strategies utilized
5. System for evaluating student learning and teacher effectiveness (impact) (Suffolk's Virginia is for Learners Innovation 3.0 Team Artifact)

We set our sights on how we could go back to basics with teachers and administrators and rebuild the capacity of teachers to engage in effective instructional practices in the classroom everyday.

In May of 2022, the C & I Team held a retreat during which we focused our efforts on effective instructional strategies, the need for professional learning communities, and brainstorming how all of this could come together for administrators and teachers in the 2022-2023 school year. During the retreat, the C & I team developed instructional models for grades K-12 to provide a clear structure for allocated instructional time at each level. The team drafted models by content area and provided each other feedback to ensure that it was clear for teachers. Additionally, the group collaborated with the instructional technology resource teachers to identify at least three critical instructional strategies that were evidence-based and included a technology component so we could continue with the technology emphasis begun during COVID.

Also, during the C & I retreat, content teams reviewed data points and screeners to ensure that we provided a clear method for teachers to collect and analyze data. This culminated in the development of a common professional learning community protocol document that guided teacher teams in how to effectively analyze that data to make instructional decisions in the best interest of students. This day of collaboration sparked the idea for us to begin this "Back to Basics" approach with school administrators.

Tier-1 Instruction Boot Camp

In *The Power of Doing Less in Schools (2022)* Justin Reich states, "If the system has to be fixed, and we can't fix the system by adding to it, then the logical place to start is with *subtraction*." (23) The SPS C & I Team recognized that COVID-era instruction had placed teachers and administrators in an awkward and ineffective space. We determined that we would unwind them from that space through a Tier I Instruction Boot Camp beginning summer of 2022. According to The American Institutes for Research (2021), Tier I instruction must include evidence-based practices aligned with grade-level skills that allow students to make connections across content areas while using explicit instruction along with behavioral and targeted academic support. Our Tier I Instruction Boot Camp was designed to provide administrators with the foundational skills needed to revisit what good teaching looked like.

In the summer of 2022, the C & I team facilitated full-day training for school administrators on planning, teaching, assessing, and creating professional learning communities with teachers. The training did not include sit-and-get instruction. In early 2021, SPS had recently adopted a new 5E Innovation Lesson Plan template that prompted teachers to consider the learning cycle as they planned instruction. We actively engaged school administrators working through the 5E learning cycle as they traveled through all areas of effective instruction. The 5E learning cycle, designed by Roger Bybee for the Biological Science Curriculum Study (2009), outlines how students learn through engagement, exploration, explanation, elaboration, and evaluation. School administrators participating in Tier I Instruction Boot Camp were taken through each part of the learning cycle as they learned, prompting them to better understand how students gain skills and knowledge daily.

The focus on Tier I instruction was designed to not be a "new" initiative, but a return to the foundations of our T & L Framework and what works for students. The unwinding of the COVID-19 instruction required staff to revisit how students learn and what helps them engage with learning in a way that is meaningful to them. All content areas identified strategies and non-negotiables that would help make learning more effective. The SPS C & I Team developed instructional models for elementary, middle, and high school to demonstrate how to most effectively use the block of time allocated to each subject or course. These models were rooted in child development and the 5E learning cycle; the emphasis was on more doing and less getting by students. These models outlined the suggested time to be spent in each of the 5Es and what evidence-based practices or activities should be conducted. See Figures 1.3 and 1.4

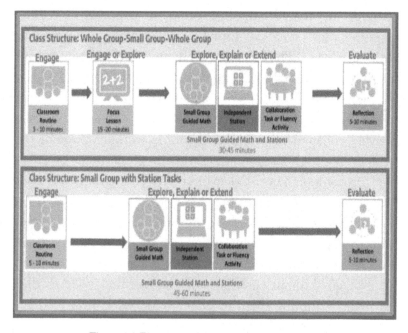

Figure 1.3 Elementary Math Instructional Model

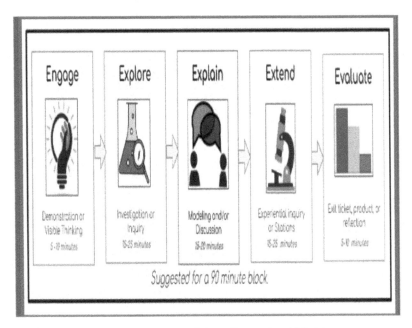

Figure 1.4 Secondary Science Instructional Model

Once school administrators had completed the boot camp, the C & I Team prepared to train teachers during preservice in August 2022. The idea of providing professional development first to administrators and then teachers, was something the team felt would best serve all students. During preservice, each content area identified its nonnegotiable tier I strategies and provided teachers the opportunities to engage with those strategies prior to the start of the school year. Additionally, schools were tasked with setting up professional learning communities among grade levels and courses so that teachers could begin to build collective efficacy around their planning and instruction. Teachers were trained and provided the Collaborative Learning Teams Protocol Template designed at the C & I retreat.

Figure 1.5 Collaborative Learning Team Meeting
Template

SPS had set the T & L Framework as the expectation for all classrooms to best serve students and help them move beyond the learning gaps generated by COVID-19 closures.

Tier I and STEM Revisions

As SPS found itself with lower-than-expected scores in science following the administration of the 2021--2022 Standards of Learning Exams, we looked to Tier I instruction as the foundation for rebuilding teacher capacity in science.

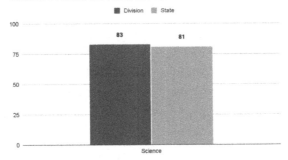

Figure 1.6 Suffolk Public Schools Science
Performance 2018-2019

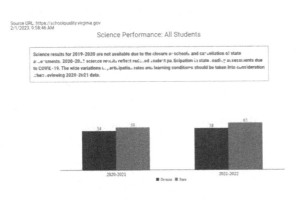

Figure 1.7 Suffolk Public Schools Science
Performance 2020-2021 and 2021-2022

The division funded science lab materials and required all teachers to conduct at least one lab per quarter with students. We continued to provide project-based assessments in grades K-12 and provide common formative assessments to ensure that we could monitor student progress as it was happening. Project-based assessments require students to design a project with a focused objective based on a specific set of subject-specific skills or standards. The assessments ask students to demonstrate their understanding of these skills by designing a student-centered project. Common formative assessments are short assessments designed to measure a student's mastery of specific skills and standards after a unit or topic has been taught. As a division, we created common formative assessments (CFAs) for all core content areas - English, math, social studies, and science to provide teachers with relevant and timely data to drive instruction. We established learning communities among groups of teachers to build teacher efficacy and allocate time for teachers to examine the common formative assessment data to inform day-to-day instruction. We provided safety training to all teachers to ensure they had the skills to conduct labs safely and without incident. Our science coordinator plans regularly with our lowest-performing schools to build that capacity among new school teams, some of which include long-term substitute teachers.

In addition to a focus on science, the division continued this model of Tier- I instruction in the area of mathematics. The SPS Math Department implemented high-yield strategies such as three act tasks, fact fluency, number sense routines, and concrete and representational models to support authentic math learning. The math coordinator, Kelly Greening, provides consistent professional development on these strategies to academic coaches, paraprofessionals, classroom teachers, and special education teachers. We use professional development to target the areas of weakness based on the common formative and quick check assessments. For example, one area of weakness was the ability of students to solve technology-enhanced items (TEIs) with accuracy. The Math Department addressed this weakness by providing targeted training on specific high-yield strategies like "WODB-Which One Doesn't Belong?" This strategy requires students to engage in discussions to assist them in mathematical thinking, which is needed to solve more open-ended math questions like TEIs. The use of these assessments allows school administrators, teachers, and the C & I Team to monitor students' progress and provide focused support more quickly in the greatest areas of weakness.

Next Steps

As we prepare our students for 21st century learning, we are equipping our teachers to support student learning with hands-on strategies to engage students in STEM skills to help them attain the 5Cs. The process and the action steps we have taken to return to what works in learning can easily be seen in our STEM classrooms and will be visible through the academic growth of our students in the future.

∼

References:

American Institutes for Research. (2021). Tips for Intensifying Instruction at Tier I. Retrieved

from   https://mtss4success.org/sites/default/files/2021-08/Tips_Intensifying_Instruction_Tier_1.pdf

Bybee, R. (2009). (rep.). *The BSCS 5E Instructional Model and 21st Century Skills*. Retrieved

from   https://sites.nationalacademies.org/cs/groups/dbass esite/documents/webpage/dbasse_073327.pdf

Etchells, M., Brannen, L., Donop, J., Bielefeldt, J., Singer, E., Moorhead, E., & Walderon, T.

(2021). *Synchronous teaching and asynchronous trauma: Exploring teacher trauma in the wake of Covid-19. Social Sciences & Humanities Open.* 4: https://doi.org/10.1016/j.ssaho.2021.100197.

Ozamiz-Etxebarria, N., Santxo, N., Mondragon, N., & Santamaria, M.(2021). *The Psychological*

State of Teachers During Covid-19 Crisis: The Challenge of Returning to Face-to-Face Teaching. Frontiers in Psychology. 11:620718. doi:10.3389/fpsyg.2020.620718.

Reich, J. (2022). The Power of Doing Less in Schools. *Educational Leadership, 80*(2), 22-27.

About the Author

DR. MARIA LAWSON-DAVENPORT

Dr. Maria Lawson-Davenport currently serves as the Director of Curriculum and Instruction for Suffolk Public Schools.

She has experience teaching middle and high school, correctional education, adult basic education, and community college.

She received her Bachelor of Arts in English from Hampton University, her Master's of Education in Secondary Curriculum and Instruction from the College of William and Mary, and a Doctorate of Education in Adult Education and Program Development from Regent University.

THREE

STEM Exploration

Mrs. Catherine Pichon

"Exploration is a wonderful way to open our eyes to the world, and to truly see that impossible is just a word." —Richard Branson

---

LIKE MOST CHILDREN, AS SOON AS MY SON STARTED WALKING and talking, he was on a mission to discover the world. I think *"why"* was his favorite word at that time. As educators and parents, we are responsible for encouraging children's natural curiosity and wonder about the world around them (Master, 2017). This is why STEM exploration is so important. Through questioning, critical thinking, experimenting, data collection, and problem-solving, children develop skills that will serve them in many ways in the future (McClure, 2017).

As Director of Elementary Leadership, I get a close-up view of how Suffolk Public Schools encourages elementary students and teachers to explore STEM. Our content lesson plans include opportunities to engage, explore, explain, elaborate, and evaluate.

Engage Anticipatory Set	TTW Models Friction Demonstration. 1. Blow up the balloons and tie a light string to each 2. Hold a balloon by the string (it should be hanging down) and bring the balloon close to each of the materials (the second balloon, the tissue paper, and the aluminum can). Observe what happens next to the second balloon, next to the tissue paper, next to the aluminum can. 3. Rub both balloons onto your hair or onto the wool fabric. 4. Hold a balloon by the string and bring the balloon close to each of the materials. Observe what happens. You may have to rub the balloon on your hair or wool again after a few minutes. 5. Bring the balloon close to a stream of running water from your kitchen sink. Make observations. TTW ask students, "What can you conclude after looking at your data?" __Teacher Notes:__ Rubbing the balloon onto your hair or onto the wool fabric adds electrons to the balloon and causes the balloon to become negatively charged. Like charges repel (the two balloons, once charged, will move away from each other) and opposite charges attract (the paper will be attracted to the charged balloons.) Your positively charged hair is attracted to the negatively charged balloon and starts to rise up to meet it. Similarly, the aluminum can is attracted to the negatively charged balloon.
Explore Student-Led Inquiry	TSW complete the _Dragon Racer Investigation._ TTW need to print the cutout on the page 4 of the investigation.
Explain Teacher Modeling & Guided Student Practice	TSW share their investigation and results. The class will graph each group's result. TTW explain friction is a force that opposes the motion of an object. TTW use _Friction Chapter in DE Techbook._ __Resource- Discovery Education:__ _The Pulas Friction_
Extend Independent Practice	_Discovery Education Friction Simulations_
Evaluate Closure	TSW can use the reflection questions from the investigation as part of the evaluate.

Our teaching and learning department provides STEM lesson plans with the materials needed to conduct them for every grade level each quarter. Our locally developed performance-based assessments (PBAs) are STEM-focused, and students complete one PBA each quarter. Our elementary schedule includes a morning meeting block during the first fifteen minutes of the instructional day, which often includes open-ended questions and opportunities for exploration. The typical components of morning meetings include greetings, opportunities to share, activities that build classroom community, and a message to begin the day.

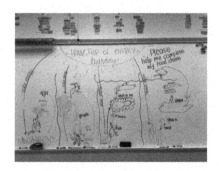

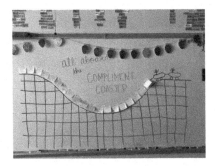

We incorporate STEM exploration within the enrichment portion of our summer series program. We also provide opportunities for families to engage in STEM exploration during each school's monthly family engagement evening events and over the summer through our SPS Explore program

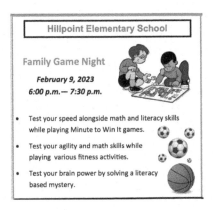

Additionally, we have several partnerships like the Nansemond River Preservation Alliance, the Virginia Air and Space Center, and the Virginia Living Museum that allow our students to engage in STEM and career exploration. The Nansemond River Preservation Alliance visits our schools and does an in-service with third-grade students to showcase the STEM boxes they will engage with. They bring critters and oysters for students to explore to get a starting point on their watersheds exploration STEM boxes. They also developed a student packet with videos locally in Suffolk for students to use when investigating our watershed. The Virginia Air and Space Center provides virtual and outreach opportunities to all fifth-grade classes. These opportunities are aligned with our Virginia Standards of Learning and incorporate STEM career connections. They also offer field trips to select schools, and it is all free of charge to our division. The Virginia Living Museum provides a way for families to engage with and learn about wildlife and animals within our region.

STEM exploration does not end in the classroom! Our elementary schools have outdoor gardens too! Students have access to various outdoor learning opportunities tied to STEM and social skills. We also provide field trip experiences for our students beyond the classroom. Imagine the possibilities when students have access to real-world learning experiences at farms, zoos, museums, and planetariums!

In my role, I also serve as the district's Early Start Coordinator, and boy, can I attest to the fact that our four-year-olds are engaged in a lot of STEM exploration! Our STREAMin3 curriculum has built-in STEM activities. STEAM (Science, Technology, Reading, Engineering, Art, Mathematics) in3 (integrated, intentional, interactions) is an engaging curriculum for children ages birth to five developed at the University of Virginia. It focuses on exploration, investigation, problem-solving, and discovery while supporting STREAM skills. Teachers also provide a variety of materials during center stations and small group instruction that promote creative thinking and allow for exploration. For example, students in the block center work together to solve problems collaboratively using math and scientific investigation concepts. With some initial guidance, teachers ask open-ended questions to guide students' thinking and promote creativity. Our early-start students are becoming problem-solving scientists in the process!

Research demonstrates how important it is for children to acquire early STEM knowledge and skills and how critical it is to support educators in teaching early STEM (Lange, 2019). What young children learn about math and science before kindergarten can have a lasting impact on their future academic performance (Lange, 2019). Not only do researchers understand the importance of integrating STEM and early childhood education, but Suffolk Public School's early childhood educators also understand the importance. They shared their thoughts on why STEM is important in early childhood education at a recent monthly meeting:

- STEM allows students to experience the freedom to explore their environment and learn about real-life concepts.

- It provides excellent hands-on learning opportunities that integrate different content areas.
- It opens doors to every area of development and encourages risk-taking and analytical skills.
- It drives students to utilize and strengthen cognitive, communication, fine motor, and interpersonal skills to become confident problem solvers.
- It allows students to see an idea in action and use higher reasoning to solve problems to make something work (like our roller coaster SPS STEM project).
- STEM allows children to solve problems and work together, building social and
- collaboration skills; it can also offer opportunities to develop a higher vocabulary level.

A STEM education is accessible to young children everywhere, as long as we give them the space and time to explore!

STEM emphasizes the types of problems people are paid to solve every day; thus, it helps bridge the gap between school and the workplace (Gallagher, 2019). It is my responsibility to encourage, support, and facilitate STEM exploration in order to help bridge the gap between elementary and middle school so that our students will enter middle

school as confident, curious critical thinkers and problem solvers. I navigate this responsibility by coaching leaders through modeling and cheerleading! Coaching is all about supporting high-quality instruction! Modeling takes place during collaboration with leaders on problems of practice and engaging them in professional learning to keep them abreast of instructional strategies and resources that support STEM exploration. Cheerleading involves encouraging risk-taking and being present! This coaching has a ripple effect from the director, to the school leaders, to the teachers, to the students. Imagine the possibilities from these two scenarios:

A teacher has an idea for the blacktop space next to the school building. He shares his idea with the principal, who in turn has him share it with the Director of Elementary Leadership and the Director of Career and Technical Education (CTE). The group brainstorms ways to fund and execute the idea utilizing division resources and high school students in the CTE program. A week later, the Director of Elementary Leadership shares a Day of Action flyer with the teacher and principal. United Way's Day of Action involves a group of volunteers supplying the manpower, and sometimes some funds, to help schools complete needed projects. The school is currently putting the finishing design elements on this project, and their outdoor learning pavilion will be built soon with the help of community partners and the amazing students and staff of Suffolk Public Schools.

Two fourth grade girls, after reading a book, become inspired to start a club. They write a proposal and share it with their school princi-pal. The principal lets them run with their idea, and the director is invited to the first club meeting as a guest speaker on the topic of leadership.

Nothing is impossible when curiosity, creativity, collaboration, and leadership are encouraged and supported.

I quoted Richard Branson in the epigraph at the beginning of this chapter. It is only fitting that we end with a line from one of his most famous quotes that exemplifies the mindset needed for STEM exploration advocacy and why STEM exploration is a catalyst for student achievement, confidence, and social-emotional wellbeing, "You learn by doing, and by falling over. Do not be embarrassed by your failures; learn from them, and start again."

~

R eferences:

Gallagher, M. (2019, June 7). The What, Why, and How of STEM in Elementary Education. *KidSPARK Education.*

https://blog.kidsparkeducation.org/blog/what-is-stem-education-and-how-do-i-teach-it-in-elementary-school

Lange, A. (2019, October 14). Engaging Preschoolers in STEM: It's Easier Than You Think!

https://dreme.stanford.edu/news/engaging-preschoolers-stem-it-s-easier-you-think

Master, A. (2017, March 31). Make STEM Social to Motivate Preschoolers. *NAEYC.* https://www.naeyc.org/resources/blog/make-stem-social

McClure, E. (2017). More Than a Foundation: Young Children Are Capable STEM Learners. *Young Children*, 72(5).

https://www.naeyc.org/resources/pubs/yc/nov2017/ STEM-learners

About The Author

MRS. CATHERINE PICHON

Catherine Pichon, Ed.S has 23 years of experience in public education as a teacher, academic coach, assistant principal, principal, and director. She has been honored as Reading Teacher of the Year and Teacher of the Year and has presented at several conferences on teaching for equity, school leadership, and school improvement. She is pursuing a doctoral degree in Advanced Educational Leadership at Regent University, serving as the proud Director of Elementary Leadership for Suffolk Public Schools, and fulfilling her most coveted roles as wife and mother.

FOUR

STEM: The Early Years

Mrs. Ashley Nettles

IF YOU WERE TO WALK BY MY 5TH-GRADE CLASSROOM DURING a STEM activity, you may be surprised to hear how loud the room is from talking and laughter. To the average person, my class probably sounds chaotic, and sometimes I would have to agree with that observation. I like to say that STEM instruction in my classroom is organized chaos. While on the outside, it may seem disorganized, on the inside, the loud talking is rich conversation and the laughter you'd hear is from our excitement and success. I have found that the days we are using STEM in the classroom are my most fulfilling days as a teacher. Unfortunately, my class has not always looked that way this year. Getting to the point of organized chaos required a shift in my own thinking. It required me to realize that good instruction is not always in my control. Good instruction could mean my elementary students are getting messy, talking to others, and experiencing the excitement and frustration of trial and error. Luckily, I have a great group of students that inspired me to make this mindset shift.

Before I started to teach content through STEM, my classroom was silent during our science block, and if I'm being honest, it was boring. I was bored, and I knew my students were bored too. I caught myself in a cycle of giving notes and teaching to the test which was exactly what my college professors told me not to do. While I knew this was not the key to great instruction, I knew that teachers at any level have a profound amount of pressure on them when it comes to student success. In my own mind, I could control that pressure by making sure I gave the students what they needed rather than letting them experience these Science skills themselves.

During the science block, I would do demonstrations with STEM materials, but I was the one doing STEM activities, not my students. For example, I did a demonstration showing how we could filter muddy water through a coffee filter to separate the solution into clean water and the leftover dirt. At the moment students were engaged and excited to see the demonstration, but in the long run, they were struggling to understand how it related to the content. While I was exposing my students to the content, I began to feel frustrated as I would watch the scores from each test come in, and my students were not doing as well as I knew they could be doing. I realized I had to take a step back and reevaluate how I was teaching my students. It dawned on me that these were elementary school students and they needed fun and excitement. They did not need or want a teacher that constantly gave them notes. They wanted to be scientists, not watch me be a scientist. Thinking back to when I was in the 5th grade, the activities that involved experiments or technology were my favorite. To this day I do not remember the information I learned from completing worksheets in the 5th grade, but I do remember the hands-on learning experiences my teachers gave to me. I knew I needed to make a change in my classroom, and that change had to start with me.

This change in mindset required help from those around me. Luckily, I knew I had a lot of support from colleagues both in my building and downtown in our School Administration Office. After working with our science leaders in the School Administrative Office I gained the confidence to do hands-on activities daily to help my students with the science skills we were working on. The idea of doing STEM every day sounded exciting and terrifying all at the same time. I noticed my students were more engaged in our lessons, and it was safe to say that we were all having more fun.

During our Earth Surfaces unit, we used STEM instruction to create and model the various concepts we needed to know. Students created the layers of the Earth with modeling clay as well as terrariums that were filled with layers of rice dyed to reflect the different layers of our Earth. My class also explored how our plate tectonics could move by using whipped cream and graham crackers. Each day I was excited to have new materials on display for students to see when they arrived to class.

Without fail, upon arrival, one particular student would make it his mission to guess what the materials could be used for. This was a student who, up until this point in the year, never passed a science assessment. That always baffled me, considering he would say science was his favorite subject. The only problem was, I wasn't teaching science in the way that he needed to be taught. He was one of my students with autism that thrived with hands-on learning experiences. When I made the shift to STEM in our classroom, he began to contribute more to class discussions, and it was obvious from his smile and laughter that he was enjoying himself. He wasn't just learning about science; he was doing science.

He was particularly fascinated by the Earth Surface unit that we were working through. When I asked him what about the unit that grabbed his interest, he simply said, "Well, we live here." I couldn't help but laugh because it was such an off-the-cuff response, but he was right. My students were finding value in these STEM activities because they were meaningful to their lives here on this planet. They were able to relate their own experiences back to our instruction, something that definitely couldn't be done by just taking notes.

When the time came for our unit test, I admittedly was nervous. I wasn't the one taking this assessment, but I still held that responsibility for student success close to my heart. We began the test like we normally do, with positive affirmations and then passing out materials. My friend in class was particularly confident because he had enjoyed all of the experiments and knew he was ready. He was practically singing our affirmations because he was so ready, "I am smart, I am capable, I can do hard things!" I, however, was still a nervous wreck. When the time came for him to submit his assessment, he brought me his paper and asked if he could go ahead and submit it. This paper was something we call a "Brain Dump." I tell the kids to get out all of the information they can remember when they start the test so that they can use it as a reference sheet. Now this friend strongly dislikes taking notes, so he normally just turned in a blank brain dump, but not on this day. When he handed me his paper, my jaw dropped to the floor because what I saw was the most beautiful illustration of everything we had covered in our STEM lessons throughout the week. Every time I glanced at his paper, I would notice something new. I am pretty sure he thought I had lost my mind when I was just grinning at that paper. He finally asked again, "So, can I submit?" I just remember nodding my head really fast.

When I heard the chime on my computer alert me to a test submission, I definitely held my breath. Before I could look at my own computer I saw him clapping at his seat when he saw his own score, an 86%. I wish I could have bottled up the excitement we both had! Up until that point, his average  test score in science was 34%, which is a low F. At that moment, I knew that making the shift to STEM instruction for all of my students was the best thing for us as a class and without the support from my colleagues and inspiration from my students, this success would not have been possible.

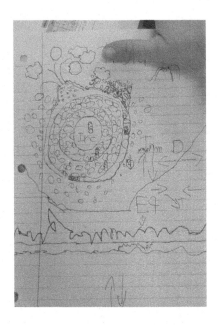

About the Author

MRS. ASHLEY NETTLES

Ashley Nettles was born and raised in Suffolk and attended Suffolk Public Schools. After graduating from Lakeland High School in 2010, Ashley went on to earn her Bachelor of Science and Master of Education from Old Dominion University. She is currently in her eighth year of teaching, and al eight years have been in Suffolk Public Schools in 4th or 5th grade. In the 2020-2021 school year, Ashley was awarded the City Wide Elementary Teacher of the Year.

Section III - The Practitioners - Workforce Identification and Specification

ONE

STEM Economic and Workforce Development

Mr. André Skinner

CAREER AND TECHNICAL EDUCATION (CTE) PROGRAMS HAVE a major impact on the role of successfully transitioning Suffolk Public School students to postsecondary education and the workforce. Our CTE programs provide opportunities for students to apply learned technical and career readiness skills and core academic knowledge through relevant work-based learning experiences. In an effort to increase the rigor of our CTE programming, we were required to integrate academia with career-based education. This integration creates well-rounded students with a refined skill set who more readily prepared to become gainfully employed. The academic and technical skills training within our CTE curriculum consists of 21st-century task-based competencies developed by the Virginia Department of Education (VDOE). These task competencies encompass work-based readiness skills as well as additional STEM-based tasks involving critical thinking and collaborative processes. During the school year, the CTE department collaborated with science teachers and Dr. Katelyn Leitner, STEM Coordinator, to create intentional professional development on the 5E learning cycle. The 5E learning cycle is important for CTE teachers to understand how to focus on teaching concepts to students through a series of five phases of continuous learning: Engage, Explore, Explain, Extend, and Evaluate. CTE teachers met during city-wide meetings and actively participated in discussions and the creation and alignment of innovative lesson plans that facilitate 21st-century learning. To ensure we are preparing students for tomorrow's careers, our teachers collaborate with local business and industry representatives to gain feedback on our programs' improvements. One of our school partners that is deeply entrenched within this process with SPS is AMADAS Industries, the world's leading manufacturer of advanced harvest systems for peanuts. They have provided both program and student feedback that was responsible for creating our SPS/AMADAS school-to-work internship program. This effort introduced students in our welding program to improve communication, collaboration, and creativity in the workplace along with additional 21st-century skills necessary for increased productivity in the workforce.

Upon completion of course task competencies, students have opportunities to showcase their mastery of these tasks by successfully passing industry credentialing tests. A credential is a verification process that evaluates one's qualifications and is awarded through a certificate, certification, license, or degree from a post-secondary institution. The focus of Suffolk Public Schools is on helping our students to acquire industry certifications within our secondary school environment. As a result, for the 2021-2022 school year, students enrolled in CTE courses earned a division total of 2,317 industry certifications. The attainment of these certifications serves as a predictor for success and oftentimes a prerequisite necessary for our students to obtain interviews for future careers.

Students are considered completers when they successfully pass two sequential electives from the same career pathway that build on related skill sets. For the 2021-2022 school year, Suffolk Public Schools attained 612 total completers resulting in a 96% pass rate on industry credentialing assessments. Completers that attend programs at our division career and technical education center, The College & Career Academy at Pruden (CCAP), receive an annual completer ceremony and are awarded a certification of completion. Due to our students' limited work experience, the certificate of completion and an industry certification serve as a resume builder.

Students in grades 10-12 from our three high schools are able to enroll at CCAP. CCAP students have an alternating schedule where they report to the first block class at their base school, then are transported every other day for five hours to be engaged in their specialty program. On days students do not report to CCAP, they remain at their base school for the core academic classes (e.g. math, English, science, history.) Whether students are interested in pursuing a college degree, industry certification, career enhancement, or life-long learning, enrolling in a CCAP program is a valuable first step for students in preparation for the careers of their choice.

Jobs in the CTE field today also require strong reading ability, communication skills, and varying degrees of math and science. As we transition our students into the workforce, CTE programs are tasked with ensuring a work-based learning focus, effective pedagogy, and relevant high-quality work-based learning experiences are readily available. High-quality work-based learning consists of twelve methods of instruction that the state department groups into three categories: Career Awareness, Career Exploration, and Career Preparation. Each phase has become entrenched in the fabric of Suffolk Public Schools CTE programs. Within these three categories, Suffolk Public Schools has chosen to focus on five instructional methods to expedite student success. These methods include service learning, school-based enterprise, internship, clinical experience, and cooperative education.

In order to give students a practical, progressive learning experience and to address social problems, service-learning is a teaching strategy that blends educational goals with volunteer work that should positively affect the surrounding community. School-based enterprise is an opportunity for students to showcase their work-based learning experiences while being employed within their base schools, removing the transportation barrier. These student-run businesses are created by partnerships with the surrounding business community and are supervised by CTE faculty. Internships and/or clinical experiences require partnerships with businesses that allow students the opportunity to acquire job experiences within their work environment. These opportunities can be non-paid or paid depending on the businesses. For example, if a student is enrolled in our Health Services programs they complete their work experience within a healthcare facility. Lastly, cooperative education combines theoretical classroom instruction with periods of hands-on, practical experience in a work environment monitored weekly by CTE teachers. Students alternate between full-time jobs and academic study through the co-op program, acquiring experience in their chosen field of study.

We developed career awareness activities to help students become more aware of their own interests and skills as well as the education and training required to pursue a career goal. Our students obtain a basic understanding of employment, numerous industries, and various career pathways through career awareness activities. As students begin to put into practice the skills they acquired in class, career awareness activities help them gain a deeper understanding of possible career paths. Suffolk Public Schools delivers the state's Career Investigations course on the middle school level to all 6th-grade students to complete inside their elective rotation. Some Career Investigations course activities include: guest speakers from Chick-fil-a, Veterinary Clinics, Huntington Ingalls; field trips to our division's technical center; or presentations about various occupations. As a team, the division felt it necessary to expose students at an earlier age; therefore, we incoporated task competencies from a Career Exploration course into fourth and fifth grade computer science resource classes, with collaboration from Dr. Katelyn Leitner, STEM/Computer Science Coordinator. Within this computer science course, students explore Virginia Wizard to identify their strengths and potential career interests. The computer science facilitator collaborates with school counselors to provide students with knowledge on available CTE courses at the middle school level. Our goal is to expose students early to CTE course offerings to assist them in their transition to secondary education.

To make decisions regarding secondary and postsecondary education and training, career exploration opportunities support the development of workplace readiness skills with a greater understanding of the career clusters available to students. The 17 career clusters recognized by the state department are: Agriculture, Architecture and Construction, Arts/AV Technology and Communications, Business Management and Administration, Education and Training, Energy, Finance, Government and Public Administration, Health Science, Hospitality and Tourism, Human Services,

Information Technology, Law/Public Safety/Corrections and Security, Manufacturing, Marketing, STEM, and Transportation/Distribution and Logistics. Suffolk Public Schools currently offers a variety of courses that are embedded within these identified career clusters. Included in these programs are more specific, hands-on training courses that are located inside our division's technical education center, The College & Career Academy at Pruden. Students who attend CCAP have schedules at their home school that include Math, Science, English, and History courses while also ensuring they have alternating days to attend CCAP for hands-on training and lab sessions for their selective careers. Students also have an opportunity to explore a variety of courses at the middle school level with the ability to adjust their CTE pathway as they update their academic and career plans with parents and school counselors annually.

Career preparation is where student's deepen their knowledge and develop the skills that are essential for success in the workplace and in postsecondary education. These opportunities are advised for students who know exactly what they want to do after high school, whether that is to start working right away or enroll in a closely linked postsecondary training program. These opportunities are set up largely to provide students with ample amounts of practice using the essential knowledge and skills for the careers they have chosen.

Within our CTE courses, career preparation occurs in numerous ways. Students engage in project-based learning and laboratory sessions that mimic day-to-day routines in the world of work. Another career preparation opportunity is the King's Fork High School Credit Union sponsored by our local financial partner in education, BayPort Credit Union. The credit union has been open for five years and currently has 39 participating accounts. Students are attempting to re-establish more student accounts after the COVID-19 pandemic school closure. This student-run school-based enterprise provides opportunities for students to engage with the student body as well as instructional and non-instructional high school staff in a professional environment. Students are responsible for opening new savings accounts, depositing and withdrawing money, and reconciling accounts under our CTE staff's guidance and supervision. Lastly, our culinary arts students at CCAP had the opportunity to engage and learn from Chef Malcolm Mitchell, chef, author, and restaurateur who starred on the Food Network. Additionally, he was the featured chef for the Denver, Colorado Democratic National Convention for President Barack Obama. During this fine dining event, students prepared a four-course meal for staff members to enjoy. Students learned the importance of table setting, meal preparation, customer service, and secrets of the culinary trade. Students were enamored with Chef Malcolm's ability to alter the taste of several foods they prepared for the event. As one student marveled, "I don't understand how he was able to make the lamb taste like beef." Chef Malcolm was then able to explain that braising the lamb, instead of simply roasting it, altered the taste for this particular menu. Experiences and mentorship opportunities such as these allow students to test their acquired knowledge gained from the classroom into actual practice.

After navigating through our CTE coursework and programming, obtaining credentials to verify comprehension of specific skills sets, and graduating from high school, many of our students still lacked the opportunity to attain high-quality and relevant work-based learning experiences. Our objective is to prepare students for the jobs of today and tomorrow. In support of this effort,, SPS monitors the labor market and employment data to align and guide students toward the selection of high-demand/high-skill career choices. During the last five years the Information Technology Cluster has shown a positive trend in employment projections in several related career pathways, thus prompting SPS to submit applications to the Virginia Department of Education to approve three new courses available within the Information Technology cluster. Cybersecurity Systems Technology, Cybersecurity Operations, and Game Design and Development courses were approved, and students were enrolled into courses to produce future employees in the field.

As we continue to grow our CTE programs, we still feel that our students need proven methods to assist with transitioning students into the workforce, thus eliminating STEM career projection shortages in the employment pipeline. Suffolk Public Schools has taken on this challenge and will be launching, summer of 2023, the Start Today And Rise (S.T.A.R.) Summer Internship Program for students who want opportunities to participate in high-quality work-based learning experiences. These opportunities will reinforce and strengthen classroom learning and prepare students for individual responsibility, teamwork, and leadership in their chosen career pathways. This program targets rising juniors and seniors who are enrolled in sequential CTE elective courses with identified career pathways. Our students will submit applications and utilize the skills obtained during their interview process to participate in our homegrown summer internships. Students selected will participate in a nine-week internship program within the athletic, facilities and planning, finance, food and nutrition services, human resources, technology, and transportation departments while earning $14.46 per hour, 20 hours a week. Suffolk Public Schools has decided to embark on this journey because it provides our students relevant work experience while also sending a message to our business community that we want to be the largest employer of former Suffolk students. Providing students with experiences in the S.T.A.R. program will provide a succession plan with a pool of qualified student support staff with the skills necessary to potentially advance to the next level of their career within Suffolk Public Schools.

In addition to our division employing our own students for work-based learning opportunities, we are thankful for our local partners who offer additional opportunities to SPS students Students have a chance to participate in internships at Starr Motors, City of Suffolk Public Works Department, and AMADAS. We are thrilled to report that these internship opportunities for our students have led to full time jobs with benefits at these companies. The CTE Department's yearly objective is to increase the number of internship opportunities available to students with our current partners while also expanding our partnerships with future employers.

It is the objective of our CTE department to readily prepare students for the jobs of today and tomorrow. CTE is the component of high schools that connects the demands of the job market and the requirements of students to prepare them to be contributing members of society after graduation, whether they enter the industry right away or pursue further career-focused education and training (Stringfield et al., 2017). The success of our economic and workforce development pipeline strongly relies upon prioritized discussions between educational and business community leaders. Suffolk Public Schools is and will remain committed to ensuring our students have the best possible opportunities as they transition to post-secondary education and/or the workforce.

⌇

References: Stringfield, S., & Stone, J. R. (2017). The labor market imperative for CTE: Changes and challenges for the 21st century. *Peabody Journal of Education*, *92*(2), 166–179. https://doi.org/10.1080/0161956x.2017.1302209

About the Author

ANDRÉ SKINNER

André Skinner is the Director of Career & Technical Education for Suffolk Public Schools. He has served in the public education field for 23 years, 18 of those years with the Suffolk Public Schools. Over the course of his career, he has held positions as a high school special education teacher, special needs liaison for a regional technical center, assistant principal at the middle school level, principal at the elementary level, and as a director of a regional technical education center.

He earned his Bachelor's degree from Old Dominion University and earned a Master of Science in Special Education and Certificate of Advanced Graduate Studies in Administration & Leadership from Cambridge University. Mr. Skinner believes that Career and Technical Education is an equitable pathway that promotes 21st-century skills necessary for a successful transition into the workforce or postsecondary education.

TWO

STEM Interest

Dr. Ron Leigh

For Suffolk Public Schools, STEM is not just for the highest-achieving students anymore. All Suffolk Public School students entering today's economy must be prepared for the demands of the current and future job market. In today's current economy, most careers that pay above minimum wage require the STEM skills of problem-solving, critical thinking, creativity, and logic, and many jobs require advanced math and engineering design skills. Formerly a world leader in math and science achievement, the United States is losing its competitive edge. In one current study, the U.S. ranked 21 out of 23 countries in math and 17 out of 19 countries in problem-solving (Beard, 2013).

Studies suggest that the majority of graduate students studying science and technology at American universities are not even American students. Fewer secondary students in the United States master STEM content than students in other countries. According to a 2017 Pew Research Survey, students cited the following reasons for not pursuing STEM courses in high school: the courses were too hard; they were not useful for my career; and they were boring (Kennedy, Heffron & Funk, 2018). Support for STEM is critical, especially in the middle and high school years where self-perceptions are being solidified. The vast majority of U.S. students receive STEM education in traditional public school settings. Educators everywhere are struggling with how to improve STEM literacy and how to encourage more students to pursue college and careers in STEM fields.

Fifty-two percent of Americans don't pursue STEM because they perceive that it is "too hard"; and for many individuals, perception is reality. Eighty percent of students make a conscious decision by 8th grade whether they think they are good at math and science, but it's important to note this decision is based on their perceptions of math and science and not their own ability. This problem undermines America's economic productivity and global competitiveness (Gerlach, 2013).

To battle this problem, we must teach our constituents what STEM is and how its impact can improve our community. As an educator of over thirty years, with twenty-four of those as a building-level principal, I have some recommendations and opinions that I believe are essential for the success of STEM programs in Suffolk Public Schools. First, we must be able to fund STEM activities and give our students and teachers the tools to make it engaging and fun. One way to achieve this is through Project Based Learning (PBL). PBL offers our teachers a way to develop in-depth thinking while engaging students. It creates many opportunities to encourage students' creativity, nurture their interests, and meet their learning needs. PBL also encourages students to think critically as they work to solve problems. While other disciplines might give students problems with only one correct answer, PBL activities require students to go through trial and error to determine which methods work best. This can best be seen with students in our International Baccalaureate Program. In this program, senior diploma track students are given a research question, which is the basis for the project concept. Students are asked to explain or defend their research through a variety of methods, such as creating artwork, song, dance, or conducting physical experiments. In this way the student has the opportunity to bridge their knowledge of the content with their learning experience. I've had the pleasure of participating in this process for five years as the principal of this I.B. School. The process is reminiscent of a Dissertation Defense in which staff and community members in the field form a committee to ask questions about the student's project.

STEM programs may also be able to help our students overcome learning loss due to COVID-19 as well as improve student interpersonal skills. COVID-19 had required our students to work in isolation, and now that our students are back to in-person learning, we (educators) have seen student regression in their ability to communicate and in their social skills. One such project undertaken by high school students in our division is the Sea, Air, and Land (SeAL) Challenge. Led by one of our district's Data Managers, this is a joint collaboration with the Penn State Electro-Optics Center and the Office of Naval Research. In this STEM initiative, teams of students learn about engineering by designing and building robotic vehicles and payloads. Students have twelve to sixteen weeks (one semester) to design unmanned vehicles and intelligence, surveillance, and reconnaissance payloads to compete in the challenge of their choice. True to the program's name, students build submersibles to maneuver underwater in the Sea Challenge, drones to fly in the Air Challenge, and land rovers to carry out a ground-based operation. Each team is paired with an engineering mentor to guide them through the design and build process. The robotic systems are then used to compete in the challenges which mimic missions encountered by the military, national security agencies, and first responders

The SeAL Challenge objectives are threefold. The first objective is to provide students with an opportunity to tackle a difficult engineering project while in high school. The second objective is to provide students with an awareness of the tremendous technical careers in the Department of Defense and the armed forces. The third objective of the program is to help educators and  administrators implement a successful Science, Technology, Engineering, and Mathematics (STEM) program in their school's time, budget and resource constraints. The SeAL Challenge, now in its eleventh year, is free to students, school districts, and organizations.

Professional development is a key component to make sure STEM is a mainstay in our school division and not simply a passing fad. Teachers should be well versed in inquiry-based instruction. We need to make sure that our teachers are comfortable with and are knowledgeable about effective classroom management techniques. This will allow our teachers to reach our most underrepresented population of students who are often left out of STEM activities because they are perceived as not being able to work in groups without being disruptive.

STEM courses have traditionally been a barrier for our Black, Indigenous, People of Color (BIPOC) students. We must ensure that all students have access to STEM courses and activities at an early age. I believe this would increase the number of students who take Advanced Placement courses, Dual Enrollment courses, and specialty programs such as I.B. (International Baccalaureate) and Project Lead the Way. Student enrollment in these programs start in elementary school, not in high school. Suffolk Public Schools is fortunate to have two nationally recognized PLTW (Project Lead The Way) programs in the areas of Biomedical Science and Engineering.

I asked Sarah McDonald, the Program Lead Teacher of the Biomedical Science program, what makes her program appealing to high school students today. Her response was both informative and passionate. She stated, "While science can be a tedious list of concepts, we need to show the practical use of what is learned. To truly transcend the knowledge received, we must have an application piece to demonstrate that not just learning had occurred but understanding was achieved. This is where STEM fits into the Biomedical Sciences Program, a curriculum full of hands-on labs and activities to demonstrate the learning of real-world skills. Students want to see where they fit into the complicated world, possibly to help others or make positive changes, which can be seen in their choice of profession. Having the students immerse themselves in a STEM activity that challenges them to learn advanced knowledge and then apply the knowledge in a lab activity allows students to see themselves in a professional role helping others and changing the world. My experience with STEM education at the secondary level is that many students learn more from applying their knowledge than with just memorizing the facts and figures. Students are engaged on a level that will allow them to think differently. Hands-on activities boost creativity and enhance critical thinking.

Just as impressive is our PLTW Engineering program. Students in this program must complete a capstone project in which they validate a problem, design and fabricate a solution, test that solution, and present their findings. Students in this program have built a wall and floor section of a house, created a wind power lab, and built a solar power for sheds on the athletic field, all funded by grants from the Suffolk Education Foundation. Students in this program are also building leaders through BLAST, Building Leaders For Advancing Science and Technology. Within the community, they've won the Girl Scout Gold Award project for promoting STEM to young girls and volunteering at elementary schools for school STEM night.

I believe that community support for STEM projects is essential for a thriving program in our district. We see that successful sports programs are often the result of an active and engaged booster club. Millions of dollars are raised each year throughout the country for phenomenal sports facilities, many of which can be seen in our own community. This is a testament to what citizens can do when they are passionate about their students and their success on the athletic field. Imagine what would happen if our community were that engaged in instructional and STEM activities in which every student could participate.

Finally, teamwork and collaboration are essential skills that students naturally learn from participating in STEM activities. These activities create an environment that encourages students to work in groups to find the best solution. This concept is best seen in eSports. eSports are simply explained as video games that are played in an organized competitive environment. The games can be team oriented multi-player online battles, single-person shooter games, survival games, or popular recreations of actual physical sports. At their core, video games are a product of all STEM (Science, Technology, Engineering, Mathematics) concepts coming together to form an interactive medium that forces gamers to actively learn new skills and solve problems in unique situations. Teammates quickly learn to treat losses as opportunities to solve problems and improve performance, turning to a growth mindset rather than embracing the idea that success is about innate talent. In that vein, our district has taken STEM outside the classroom and invested heavily in three state-of-the-art eSports labs in each high school. Students participate on three teams, each having two seasons. Coaches were hired for each school and tryouts held for each season, all at no cost to the student-athlete. Feedback has been extremely positive from students, staff and parents alike. eSports has also allowed our schools to connect with a group of students who would not ordinarily be involved in extracurricular activities. Students compete at the state, regional, and national levels. Our students are now eligible to compete for scholarships to major universities across the country who sponsor eSports teams at the collegiate level. I hope this gives an insight into STEM activities within our district and some suggestions and recommendations moving forward that will help us become the premier school distinct in the Commonwealth of Virginia and the nation.

~

References:

Beard, K. (2013), Behind America's Decline in Math, Science, and Technology. STEM symposium on Capitol Hill discusses how to bring back America's competitive edge. *U.S. News & World Report.* *Retrieved from website:* https://www.usnews.com/news/articles/2013/11/13/behind-americas-decline-in-math-science-and-technology

Gerlach, J. (2013), Is STEM Interest Fading with Students?. *EVERFI from Blackbaud. Retrieved from website: https://everfi.com/infographic/k-12/is-stem-interest-fading-with-students/*

Kennedy, B., Hefferon, M., and Funk, C (2018), Half of Americans think young people don't pursue STEM because it is too hard. *Pew Research Center. Retrieved from website: https://www.pewresearch.org/fact-tank/2018/01/17/half-of-americans-think-young-people-dont-pursue-stem-because-it-is-too-hard/*

About the Author

DR. RON LEIGH

Dr. Ron Leigh grew up in Elizabeth City, North Carolina, the son of two educators. Initially focusing on criminal justice, he graduated from Elizabeth City State University with a degree in History Education. Attending college on an athletic and academic scholarship, he developed a love for working with students outside of the classroom. While serving as a middle school teacher, he also became a high school and collegiate coach.

Moving into administration, Ron fell in love working in turn-around schools and helping BIPOC students to achieve academically. His experiences and success working at the elementary level prompted him to work with high school students in an underachieving high school moving it to full accreditation.

Ron is an avid outdoorsman, and loves fishing, cooking, traveling, and spending time with his children. Upon receiving his doctorate, and yes he's a "HOKIE,", Ron began speaking at several conferences focused on turn-around schools and equity. His last publication, entitled, "School Facility Conditions and the Relationship Between Teacher Attitudes" focused on the impact of facilities and teacher behaviors in those schools.

THREE

STEM Applications in the Classroom-The Science Perspective

Dr. Katelyn Leitner

Suffolk Public Schools' STEM Department fosters student learning and development of 5C skills through equitable, authentic learning experiences, project-based learning, and technical applications to ensure our students are future-ready. We pride ourselves in fostering deeper learning and the development of the 5C skills (critical thinking, creative thinking, collaboration, communication, and citizenship) through the application of classroom content in authentic, real-world service-learning experiences. The Virginia Department of Education developed the "Five C's" as a set of guiding principles to assist students in meeting the requirements and expectations of contemporary society. These ideas serve as the foundation for the Profile of a Virginia Graduate and establish new expectations for classroom education. In Suffolk, we believe that when students are given the opportunity to apply their learning beyond the classroom, it will increase classroom engagement, academic achievement, and readiness for life beyond high school. In addition to the impact in the classroom, these types of experiences can influence a student's sense of purpose, self-efficacy, and their belief in their ability to influence the world around them.

Over the last three years, the Science Department has increased the learning opportunities for students in Suffolk. There has been a heavy emphasis on increasing community partners, resources, and opportunities for our students to be immersed in real-world learning experiences. After talking with local partners, including businesses and industries, and identifying our gaps from learning loss due to the COVID-19 pandemic, we recognized that we needed to provide students with more authentic learning experiences to hook their interests while also incorporating the state's required standards of learning. Students now have opportunities to interview scientists, engage in a career day to learn about various career opportunities, and enroll in career and technical education courses aligned to their skill sets and interests. Suffolk students are learning new skills, evaluating and analyzing data, engaging in experiences and discussions, and sharing their expertise with their peers. Students plan investigations and solve challenges cooperatively with others. Students successfully make their own discoveries when active, collaborative learning is focused on scientific inquiry. These inquiry-based learning opportunities include hands-on experiments and activities where students must think critically and logically to come up with explanations. This is evident in our elementary students' performance-based learning opportunities. Students are tasked with identifying problems within their school grounds to develop an actionable plan for a potential solution. The final phase in these learning experiences is to communicate their findings and plan in order to promote effective change. We did a phased approach to incorporate science and engineering best practices. We started with early elementary and worked our way up grade levels and courses. Research indicates learning is acquired best through a cycle of events called the 5E Learning Cycle. First, students are hooked into the day's learning concept with an **engagement** activity. Then, students are given manipulatives, sorts, text, or a guiding question to **explore** with a peer. This phase is vital! Students start to take ownership of their learning when they are given the opportunity to explore. This part of the learning cycle allows all students to be on an even playing field when it comes to accessing the curriculum because they all are given a common experience to connect the content. Our role as educators is to provide students opportunities to make sense of their curriculum and make sense of why this information is of important. Who cares about solvents and solutes? When we are cooking or taking medicine, it is important to understand the best process for substances to dissolve into solutions. These are the connections we help students make through the exploration period of the learning cycle.

Following the explore phase, students then **explain** what they have learned in the explore phase. Teachers are then able to address misconceptions and provide direct instruction on the day's concept or skill. It is important for students to share their thoughts and knowledge gained from the exploration phase because, as we know, students also learn from hearing from their peers. After the explain phase, teachers then provide students with an opportunity to showcase their learning with independent work, and this is the **extended, or elaborate** phase.

Lastly but certainly not least, is the **evaluate** phase. Teachers evaluate student learning, both formative and summative, throughout the entire learning cycle. Teachers in Suffolk are receiving ongoing support and training to better understand the 5E learning cycle so students are given common experiences to create meaning and to continuously assess their comprehension of the content.

In Suffolk, students DO science. As you walk into a science classroom in Suffolk Public Schools you will experience students immersed in hands-on learning. In early elementary, students are planning and conducting investigations to better understand matter, how the direction and size of force affect the motion of an object, and how plants and animals have structures that distinguish them from one another. Upper elementary students are investigating electricity and how it is transmitted and used in daily life and applying scientific ideas to design, test, and refine a device that converts energy from one form to another. In middle school, students are engaged in identifying abiotic and biotic features in local watershed, investigating practices that can reduce environmental hazards or improve land use, and conducting investigations to provide evidence that living things are made of cells, either one cell or many different numbers and types of cells. High schoolers construct models and manipulate online simulations to represent and explain how substances move across the cell membrane by osmosis, diffusion, facilitated diffusion, and active transport. Students are analyzing their water usage at school to create recommendations for water use improvement, engaged in animal dissection, and learning through real-world scenarios using a 3D anatomy table that projects actual X-rays, MRIs, and CT scans. These experiences are molding our students to be environmental stewards.

Aside from the science department, kindergarten through twelfth-grade teachers working on the science instructional resource team during the summer developed STEM innovation plans (*Table 1*) organized in the 5E learning cycle format and aligned to their standards of learning. Additionally, the science department and the curriculum development team, including science teacher leaders, developed an investigation lab sheet (*Table 2*) for students to have repetitive practice on scientific investigation embedded into their core curriculum. Teachers are equipped with the manipulatives and consumables necessary to facilitate these rigorous learning experiences.

Instructional Innovation Plan

Teacher: _____ Grade/Subject: _____
Lesson Description:
Student Product(s):

Week of	Monday	Tuesday	Wednesday	Thursday	Friday
SOL (Standard/Skill)					
Standard Skill (Bold)					
Objective					
Tier 1 Instruction (5E Learning Cycle)					
Engage (Anticipatory Set)					
Explore (Student-Led Inquiry)					
Explain					
Reduce (Teacher-Guided Practice)					
Extend (Independent Practice)					
Evaluate (Closure)					
Assessment (Formative or Summative)					
Curriculum Integration					
Bloom's Level	Remember	Understand	Apply	Analyze	Evaluate
5Cs	Collaborate	Communicate	Critical Thinking	Creative Thinking	Citizenship
Possible Misconceptions					
Material(s) & Technology					

Table 1: STEM Innovation Plan

Aside from the STEM innovation plans, elementary students enrolled in kindergarten through fourth grade are exposed to performance-based assessments every nine weeks. Students are planting, growing, and harvesting plants and vegetables in their school gardens, creating proposals to implement plans to reduce erosion, and investigating their school yard to identify questions, problems or issues that affect a natural resource in the area being studied in order to determine possible solutions. These cross-curricular opportunities encapsulate inquiry-based learning focused on the whole child. In order for students to flourish after graduation and meet the demands and expectations of modern society they need exposure and experience to deeper learning. By asking questions about the natural world and looking into phenomena, students engage in inquiry. Through this process, students learn and get a deep understanding of ideas, principles, models, and theories. At all grade levels and in all areas of science, inquiry is an essential part of the curriculum. Inquiry-based learning allows students to learn science in a manner that is reflective of how science really works. As a division, we strive to provide opportunities for students to collaborate with community members in meaningful service projects. A few of our local partners who provide students with rich and meaningful opportunities outside of the classroom are the Nansemond River Preservation Alliance (NRPA), Virginia Air and Space Center (VASC), Peanut Soil and Water Conservation District, Wolf Trap, Community Outreach Coalition, Great Dismal Swamp National Wildlife Refuge. Additionally, as a division, we ensure throughout their work in the community that students communicate their advocacy, demonstrate creativity through presentations, embody citizenship by bringing awareness to issues and regulations that relate to an organization's area of focus, and think critically about how they might meet a community need or address a problem in society.

Name: _____		Date: _____
Title of Investigation		
Independent Variable		
Dependent Variable		
Constant		
Hypothesis		
Guided Questions:		
Safety:	Adhere to Suffolk Public School's Lab Safety Agreement Form	
	**Include any specifics for this particular lab.	
Materials & Procedures:		
Data/Findings/ Calculations/Observations		

Table 2: Investigation Lab Sheet

Suffolk Public Schools implemented a computer science resource class in each of the 11 elementary schools. Kindergarten through fifth graders are engaged in a locally developed curriculum where they acquire typing skills, explore computer science standards through code.org, and develop problem-solving skills through an engineering unit. Students learn about the design thinking process as they design a prototype based on their interests and skill set. The purpose of incorporating this into the computer science resource class is to support our core content areas so students can see how their learning is holistic.

Additionally, fourth and fifth graders engage in a career explorations unit to learn about their skills and attributes that will help them in future course takings and careers. With the implementation of this course, Suffolk Public Schools now has a K-12 pathway for computer science.

Within the last three years, the Science and Career and Technical Education (CTE) departments have intentionally streamlined our courses. There are a plethora of benefits to CTE courses as they prepare students to be college and career ready. Within these courses, students work through various task competencies that build character and 21st-century skills. A few connections we've put into action are allowing students to 3D print the physical structure of DNA or the outline of a plant cell to have a visual representation for their science objectives. This has increased the cross-curricular collaboration with our secondary teachers. Another exciting opportunity we have at Lakeland High School is the connection between our Anatomy and Physiology (A&P) class and the Project Lead the Way(PLTW) Biomedical Program. Sentara Healthcare graciously donated funds for Suffok to purchase an Anatomage table for both courses to utilize. This resource enhances our A&P and PLTW program with augmented reality technology to improve student opportunities in biomedical science studies. The Anatomage Table allows high school students to dissect a human cadaver like a medical student. This virtual dissection projects images in 3D to engage students in an operating table factor. Students are working with technology that is the same device within our local hospitals. Talk about preparing them for the workforce!

In addition to the learning experiences and field trip opportunities students are engaged in during the academic year, learning does not stop after June. Suffolk Public Schools offers a summer STEM camp at John F. Kennedy Middle School with the help of our partner at the Naval Warfare Center. Last summer, students designed and constructed vinegar rockets, circuits, robots, and more. The science department is excited to partner with 21stCentEd summer 2023 for a STEM Summer Camp. As the demand for programmers, game designers, robotics engineers, and creative thinkers rises, we are providing opportunities for students to cultivate essential 21st-century life skills An additional upcoming opportunity for our middle school students is an after-school STEM camp with our local partner Jessica Johnson, at Virginia Modeling, Analysis and Simulation Center. Students will be immersed in their advanced technology while learning about Career and Technical Education (CTE) courses they can enroll in at the high school level in Suffolk. Lastly, our kindergarten through eighth-grade students enrolled in summer school are engaged in an enrichment hour that includes a locally developed STEM curriculum. We strive to immerse our students in many learning opportunities with an emphasis on STEM education to be life ready.

As we continue to evolve our curriculum, professional development, and learning opportunities to ensure we are preparing our students to become 21st-century learners, our next phase of implementation is partnering with 21stCentEd to train 20 STEM Leads, one per building, here in Suffolk. Each STEM lead will engage in a four-part professional development workshop focused on how to build a culture of STEM literacy for every learner, infuse 21st-century skills in every classroom, make learning rigorous, authentic, and connected, and use feedback and assessment to increase performance in STEM education.

As a school division, our vision is for all students to become life-long learners equipped with the skills, knowledge, and attitudes to succeed as productive citizens in a local, national, and global society. As a science department, we will continue to provide opportunities to deepen student learning through real-world connections, cross-curricular content integration, project-based learning, and performance-based assessment. What makes Suffolk Public Schools special is that everyone contributes to STEM in the classroom; from gifted resource teachers integrating STEM into the curriculum, reading specialists promoting STEM nights, to parents and partners investing in our students to ensure they have access to an equitable education. Together we are creating achievers!

About the Author

DR. KATELYN LEITNER

Dr. Katelyn Leitner serves as the Coordinator of Science and STEM (science, technology, engineering, and math) for Suffolk Public Schools. She has been teaching in the Suffolk public school system for eight years. She earned her bachelor's degree from Radford University, has a Master of Education in curriculum and instruction from Averett University, and received her doctorate in educational leadership from Liberty University.

Katelyn was previously employed as a standards of learning (SOL) coach in Montgomery County Public Schools. As the current Coordinator of STEM, one of her main responsibilities is ensuring students are engaged in authentic experiences as they develop 21st-century skills preparing them for their post-graduation path. She is motivated to improve opportunities for students while developing a rigorous curriculum that embeds employability skills as essential for both college and career options.

#DoWhat'sBestForKids

FOUR

STEM Applications in the Classroom-The Math Perspective

Mrs. Kelly Greening

IN THE EARLY EIGHTIES, MOST LITTLE GIRLS WANTED A BARBIE Dream House or a Cabbage Patch Doll for Christmas. I, on the other hand, was over the moon to open my very first microscope at age eight. I have fond memories of the rich, hands-on experiences my mother ensured I had in science. Whether traipsing down to the local creek to collect tadpoles or exploring chemistry in the kitchen, my early experiences at home primed me for a deep love of exploring science during my academic career. Unfortunately, the same cannot be said of the anchor of STEM: Mathematics.

Like many others before me, my early math education was characterized by memorizing basic facts, formulas, proce-dures, and pages of mindless practice and drill. By fourth grade, math had become a chore of calculations and anxiety. Although I had my basic facts memorized, I struggled with long division and did not understand what to do or why. By eighth grade, my math identity was at an all-time low, and I was convinced that I was not a math person and would never be capable of doing mathematics. However, that all changed during my senior year of high school when I took Math Analysis/Trigonometry with Mr. Rios.

I can still picture the twinkle in Mr. Rios' eyes when he was about to challenge us to actually think about the work we were doing in his math class. We did not just take notes and do the odd practice problems in the textbook in his class. Instead, he challenged our thinking through meaningful contexts and tasks. I vividly remember a math task, the first I had ever done in my K-12 mathematics education, in which Mr. Rios had us use trigonometry to model the motion of Tarzan swinging through the jungle to rescue his love, Jane. Mr. Rios did not give us a word problem to solve for some unknown value. Instead, we collaborated with others to model the problem using multiple representations, propose multiple solutions, and present our thinking and solutions to the entire class. He expected us to critique the mathematical reasoning of others and use feedback to improve our models. This way of engaging in math was completely foreign to me, yet wonderful and exciting at the same time. In his class, we were mathematicians. This is the positive, meaningful experience we strive for in math classes in Suffolk Public Schools.

Mathematics is not just the last letter in the STEM acronym; it serves as an anchor to all aspects of STEM education. Math is the foundation of logical thinking and reasoning critical for all STEM disciplines. Unfortunately, mathematics can be a deterrent to many of our young people from pursuing STEM-related fields. Students see being good at math as being about getting the right answer and getting that answer quickly. All too often students are not asked to understand what they are doing in math class but instead just mindlessly follow an algorithm. However, a solid foundation for STEM requires that students be critical thinkers capable of using mathematics to explain their findings, model their thinking, and make predictions. Students need to see themselves as capable doers of mathematics. My STEM story could have been finished before it started had it not been for the positive math identity and sense of agency that was instilled in me by Mr. Rios' class. In Suffolk Public Schools, we deeply believe every student is capable of succeeding in mathematics, and that belief is at the core of all we do as a math department. We are preparing our students for STEM careers and their futures by laying a solid foundation in mathematical problem-solving and reasoning.

Our deeply held beliefs about what mathematics in a STEM classroom should feel and look like are evident in our vision and mission for our department. The vision of the mathematics department of Suffolk Public Schools is to support teachers and staff in providing all students with an equitable mathematics education in which all students develop into critical mathematical thinkers capable of using mathematical problem-solving, communication, reasoning, connections, and representations. Our mission is to foster a mathematical community in which all students develop a positive mathematical identity and sense of agency through rigorous, authentic mathematical exploration that builds conceptual and procedural knowledge.

A vision and mission are just words until they are brought to life in our students' classrooms. In a joint statement on building STEM education, the National Council of Supervisors of Mathematics (NCSM) and the National Council of Teachers of Mathematics (NCTM) advocated for a meaningful, integrated STEM program that uses grade-appropriate standards and best practices (NCSM & NCTM, 2018). For mathematics instruction, the gold standard is the eight effective teaching practices described in NCTM's *Principles to Action: Ensuring Mathematical Success for All (Table 1).*

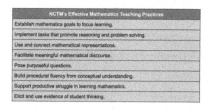

Table 1: NCTM's effective mathematics teaching practices that support quality based mathematics teaching for all students.

By developing our teacher's effective teaching practices, we believe that our students will gain problem-solving skills and a sense of curiosity and creativity critical to engaging fully in STEM education. While all of the eight effective teaching practices are equally important, we have begun shifting how our division's teachers approach mathematics instruction through a multi-year layered approach heavily focused on four of the eight effective mathematics teaching practices. They are:

- Implement tasks that promote reasoning and problem solving
- Facilitate meaningful mathematical discourse
- Build procedural fluency from conceptual understanding
- Support productive struggle in learning mathematics

Using these four practices as a lens, we have begun to reshape the traditional math classroom into a STEM Mathematics class that is student-centered with the goal to build critical thinkers and capable problem solvers. In Suffolk Public Schools, math is not a spectator sport.

The strong mathematical foundation critical for STEM success starts with number sense, first formed in elementary. Number sense is a broad term that refers to a set of skills that allow people to work with numbers and are key to doing math. Number sense involves understanding the meaning of numbers, relationships between numbers, magnitude, and operations with numbers. Students who have good number sense are able to think about numbers flexibly. Students develop number sense over time through rich experiences with math. To promote the development of number sense for our students, we have incorporated number sense routines in our elementary classrooms. In the summer of 2021, curriculum resource designers created number sense routine resources for teachers to engage their classes each day. These number sense routines are quick, low-prep daily activities that encourage students to play flexibly with numbers and concepts. Several examples of number sense routines that we have incorporated include:

- Which One Doesn't Belong?
- Same but Different,
- Number Talks/Strings, and
- Quick/Dot Images.

While in traditional math classrooms, the teacher dominates the discussion, number sense routines focus on student discourse with the teacher as the facilitator of student conversations. The routine structure invites all students to be doers of mathematics and builds a positive community of learners for all students. The success of number sense routines in our elementary classrooms led us to broader implementation in secondary mathematics classrooms during the 2022-2023 school year. The rich student discourse and building of conceptual understanding facilitated during number sense routines is one step towards building math classrooms aligned with our vision and mission.

Remember the Tarzan task from Mr. Rios' class? The rich tasks in his class engaged all students in a productive struggle while building conceptual mathematical understanding. Rich mathematical tasks engage students in making sense of a problem. These tasks often have multiple ways to approach them and do not have an obvious pathway to a singular solution, which leads to increased student discourse and creativity in the problem-solving process. In Suffolk Public Schools, we are increasing the use of rich mathematical tasks to continue to build our students' problem-solving skills. Rich mathematical tasks, provided by the Virginia Department of Education Math Department, were embedded within our district pacing guides for teachers to utilize for instruction. We have also incorporated DESMOS teacher activities aligned with our state Standards of Learning (SOLs) that promote conceptual understanding and problem-solving. It is natural for teachers to want to rescue students' thinking when it appears that students are struggling to complete a task. Along with providing resources for meaningful tasks, we have provided targeted professional learning for elementary academic coaches and secondary math teachers on promoting productive struggle in mathematics using rich tasks. Allowing all students to engage with meaningful, real-world mathematics tasks fully continues to be a goal for mathematics classrooms in Suffolk Public Schools.

COVID-19 and the resulting shift to online learning had a deep impact on math instruction in Suffolk. Hands-on exploration is a critical component of students building mathematical knowledge. Students gain a deeper understanding through actively engaging with mathematics through manipulatives and discussing and justifying their thinking with each other. However, the shift to online math instruction during COVID-19 made these best practices challenging. When we returned to full-time in-person instruction in the fall of 2021, teachers continued to primarily use digital platforms during math class instruction, and hands-on exploration was nearly non-existent. As a result, we realized that student learning was suffering tremendously. Instead of bemoaning the overuse of technology, we elected to highlight teachers who were implementing meaningful math activities such as manipulative tasks, number sense routines, and other engaging lessons with our #Mathisnotaspectatorsport Twitter hashtag and sticker initiative. Positively highlighting effective teaching practices led to increased use of hands-on, active learning tasks for students and increased student engagement and achievement.

[Pictured: Mrs. Eberhardt used hands-on learning as a teacher and continues to champion math exploration as an instructional coach.]

As we look towards the future of STEM in Suffolk Public Schools, one area of growth is the cross-curricular integration of mathematics in other STEM-related areas. One such opportunity for our students is a new data science course for high school students. Data science is a STEM field predicted to grow exponentially in the coming years. Lakeland High School, home to the award-winning Biomedical Project Lead the Way program, was selected by the Virginia Department of

Education to participate in the pilot program of this new course. In this course, students engage in data science problem-solving structure to interact with large data sets. Students formulate problems, collect and visualize data, and create predictive models based on their results. Grounded in authentic project-based learning, this integrative STEM course allows our students to gain experience with communicating effectively about data. In the course, students learn through the exploration of data. Students formulate questions and choose data sets from their local context and personal interests. They are exposed to many types of data and select tools to visualize the data and model real-world situations. Students are communicators, collaborators, and problem solvers. The course gives our Suffolk students data literacy skills to carry into their post-secondary education or the workforce. The course is now part of our math studies program and is expected to expand to all of our high schools in the coming years.

Our STEM journey at Suffolk Public Schools has just begun. As a math department, we are proud of the groundwork we have laid for our students' futures and are excited to see the progress our teachers are making to bring math to life for their students. Through the implementation of NCTM's effective teaching practices, our students will have relevant experiences to build a solid mathematical foundation critical to fully participate in STEM endeavors. Not only will our students be prepared for entry into STEM-related jobs, but they will have the critical thinking and problem-solving skills needed to be informed Suffolk citizens. We believe all of our students are capable of doing mathematics, and they, too will agree that mathematics is not a spectator sport.

R eferences:

National Council of Supervisors of Mathematics & National Council of Teachers of Mathematics. (2018). Building STEM Education on a Sound Mathematical Foundation. NCTM. https://www.nctm.org/Standards-and-Positions/Position-Statements/Building-STEM-Education-on-a-Sound-Mathematical-Foundation/

National Council of Teachers of Mathematics. (2014). Principles to actions: Ensuring mathematical success for all. Reston, VA.

About the Author

MRS. KELLY GREENING

Mrs. Kelly C. Greening currently serves as the Coordinator of Mathematics Instruction for Suffolk Public Schools.

Prior to serving as math coordinator, she served as a mathematics instructional specialist with teaching experience in middle school math and science and high school math. She received her Bachelor of Science in Biological Sciences from California State University (San Marcos) and her Master's Degree in Education Leadership (Mathematics Specialist) from George Mason University.

She currently resides in Suffolk, VA with her husband of 27 years, Mike, and their beloved dog, Jack. Mr. and Mrs. Greening are proud of their sons, Ryan and Jacob, who are both graduates of Suffolk Public Schools Project Lead the Way Engineering program and are pursuing STEM careers in physics and engineering.

FIVE

STEM in the Classroom

Mrs. Dawn Rountree

CAREER OPTIONS ARE NOT ALWAYS OBVIOUS TO PEOPLE WHEN they are young. Students are typically aware of jobs and career fields that they are exposed to in their everyday lives. This exposure comes from family, activities, and other local influences. My own influences gave me an interest in photography, woodworking, science, and space exploration. Those interests influenced my decision to take courses in high school like graphic communications, electronics, and physics. Through the inspiration of my teachers, I found a career path that suited me better than what I had told people that I wanted to be since I was in the fourth grade_–an astronaut. My teachers did not just give me a skill set, they also fostered my natural curiosity and empowered my empathy for my peers when I assisted them in understanding what we were doing in class. It was natural to follow in their footsteps.

In the mid-2000s, I became a drafting teacher at Nansemond River High School while Suffolk was in the midst of a paradigm shift in Career and Technical Education (CTE) through adjustments in instructional practices, software and technology, and course offerings. Specifically in the Technology Education program area, the classrooms received furniture, computer hardware, and software upgrades. These upgrades gave students access to industry-level software at all three high schools and opportunities for jobs through certifications in Autodesk and Adobe products. Technology Education course offerings adjusted based on enrollment, student interest, and community support. Principles of Technology 1 and 2 were phased out, and Computer Animation and Dual Enrollment Game Design and Development quickly became popular options for students. It was very exciting to be a part of a district that was open to partnering with a Community College to offer a college-level course that was not core content. We continued to show growth and change with state-level curriculum changes, course name adjustments, and relocation of courses. Technical Drawing, Architectural Drawing and Design, and Engineering Drawing and Design moved to less board drawing and began to focus more on the use of Computer-Aided Design (CAD). Technology Foundations and Technology Transfer grew through changes in the state curriculum that brought the courses to be more problem-solving and design process focused. Communication Systems and Graphic Communications changed with mainstream media and were updated to cover more content that is digitally-focused rather than print. Over time, Computer Animation was renamed Digital Visualization, Game Design and Development was relocated to a Tidewater Community College campus before making a permanent home at the Pruden Center- now the College and Career Academy at Pruden (CCAP).

Birth of Project Lead The Way in Suffolk Public Schools

Suffolk Public Schools brought the International Baccalaureate (IB) program to King's Fork High School in 2009. This prestigious program required the district to offer specialty programs at the other high schools. I was honored to be invited to participate in the specialty program steering committee. I worked with the STEM subgroup, where we investigated various possible programs that Suffolk Public Schools could start at either Lakeland or Nansemond River high schools. I was excited that Nansemond River High School was chosen as the location for a STEM specialty program.

The committee chose to propose the use of the Project Lead The Way (PLTW) curriculum instead of developing their own. PLTW offered our school system courses backed by national industry partners, high-quality equipment requirements, training for teachers, and grant opportunities to help schools start new programs. Having a seat at the table to give input about course offerings and career pathways was inspiring, and I appreciated being asked to help, but I never considered teaching the courses myself. I couldn't wait to observe the different types of courses and projects that students would be doing. I was happy teaching Graphic Communications, Architectural Drawing and Design, Digital Visualization, Engineering Drawing, Communication Systems, and Technical Drawing. Working with the other teacher in my department, we had built up our program area to the point that we needed a third teacher to teach all of the students that wanted to take our courses. It was my fifth year at the school, and I had big ideas for what Technology Education could look like at Nansemond River. I welcomed the possibility of opening our doors to a new (to us) teacher who could teach the specialized courses to the level that the committee expected. I still remember a phone call to my classroom by our Assistant Superintendent at the time when I was asked why I hadn't applied for the engineering teaching position. I told her that I didn't think I was the caliber of teacher they were looking for. Let's just say I was put in my place, and I submitted my application by that afternoon. Accepting the potential that other's saw in me was a serious turning point in my career. It was how I found my career path so it would only make sense that I needed to continue to listen to people that I respect and see as professional mentors.

As a part of the specialty program steering committee, I was given the opportunity to work with the team that designed the lab. After touring other programs around Hampton Roads, I thought about the student experience and the need for versatility in a classroom and lab. The plan was to renovate the original Manufacturing and Production Room. This room had already been repurposed before I started at Nansemond River in 2004. The Technology Foundations and Technology Transfer courses had been taught in the room, but the large manufacturing and production equipment had already been removed and we gradually moved those classes to other rooms so that two teachers were not spread among three rooms. With the knowledge that students would need traditional classroom space, access to computers, and room for hands-on projects and labs the room was divided to have a classroom and lab space. This divide would provide protection for the computers and other technology from an area where learning can get messy. Since I had been teaching the use of multiple software for five years, we took that experience and decided to line the desktop computers along the walls of the room. The layout would also provide the instructor with the ability to see student screens from most locations in the room, allowing for quick troubleshooting from across the room. My favorite design choice was that our Director of Technology agreed to install two projectors with screens that would connect to the teacher's computer. Dual projectors would give students a better line of sight for notes, videos, and software demonstrations from any desk or computer workstation. Student desks in the middle of the room would be individual work areas that could be moved to create a variety of collaborative configurations.

Classroom logistics was only one step in preparing to welcome students to this exciting new academic endeavor; training was also required for schools to teach the Project Lead the Way curriculum. AWhenI started teaching PLTW courses, teachers were students first. Each course required a two-week intensive residential training at a PLTW connected college or university. This training would cover an entire school year's worth of coursework complete with software training, class assignments, and hands-on projects. Suffolk Public Schools decided that it would be best for the instructor, now me, to complete the course training the summer before implementing it that school year to ensure that we were teaching the most up-to-date curriculum. In the summer of 2010, my class at Old Dominion University was canceled, so I sold my Dave Matthews Band tickets and drove to Penn State Burks in Reading, PA to start my life as a PLTW teacher with Introduction to Engineering Design. July 2011 brought (pregnant) me to the University of Maryland, Baltimore County for Principles of Engineering. The third year of the specialty program has students taking two courses, and this would require me to take two separate two-week-long courses in one summer. My first stop in the summer of 2012 was back to the dorms of Penn State Burks for Digital Electronics. Nightly Skype sessions with my 6-month-old made the nights a little easier to do homework. Following a short stop home with my family, I made my way to Duke University in North Carolina for Civil Engineering and Architecture. This was the perfect choice for this course because I had the opportunity to be surrounded by the beautiful architecture of the school and surrounding city. My final training in 2013 was for the capstone course- Engineering Design and Development. This course was very collaborative and allowed some time for my husband to bring my son down to Duke for a weekend visit! The opportunity to train with new teachers from different areas was a great opportunity for us to network and learn together. Each training came with its own obstacles but was rewarding by giving teachers extra insight in to possible problems that students may run into when they are working on their assignments and projects.

The most exciting part of bringing the engineering specialty program to Suffolk Public Schools was filling the lab with students interested in learning more about engineering and how it may impact their future. Our first class of students started with 19 young people from across the city ready to learn! Each year, more and more students apply to attend our rigorous program and as of the 2022-2023 school year, 307 students have made this program a part of their academic journey.

Student-Driven

Students drive what is taught in and out of the classroom. I have always felt that my classroom should reflect a similar vibe that students might find if they were actually working within that field. As a drafting teacher, I always had background music playing and would hold lessons like client meetings with assignments being given as an agenda for the week.

This allowed students to learn about time management while giving students that needed extra time a quiet opportunity to have it without anyone asking why. Due to technology restrictions, the school was still heavily focused on board drawings and AutoCAD was taught through theory and small application, Even with these hurdles, many students still selected to take the AutoCAD certification test at the end of the Engineering Drawing and passed. I later found out that some students used that certification to obtain after-school and summer jobs with local drafting firms allowing them to earn money with what they learned in my class. I approached teaching Architectural Drawing with the mindset that every student taking the class may not become architects one day but they will all become consumers of architecture at some point in their lives. This approach brought more than just how to draw a floor plan but to understanding why things are designed a certain way and how cost and repairs are impacted by certain design choices. Theming projects around student interests are always fun but I found that acknowledging their future and giving them real-world scenarios and examples had students more engaged and invested in a lesson.

Project Lead the Way provides teachers with a robust curriculum backed by industry, but teachers still have the flexibility to adjust coursework and projects that connect best with their students and community. Students take part in a design charrette during junior year. While I like many of the provided scenarios, in recent years, I have brought focus to the King's Highway Bridge and have student groups discuss and identify the benefits of a redesign of that bridge or if it should remain closed. Once students connected with the project, it was easier to have buy-in on what they would be expected to research and complete for the project. The senior capstone project is one of the most anticipated parts of the engineering specialty program. Student groups identify problems, validate them, and develop their own solutions to solve the problem. In the ten years that we have had this course, we have seen a variety of projects: carrying grocery bags, sharing an armrest at the movie theater, getting every last drop of lotion out of a bottle, a place for a sauce dip cup while in the car, to accessible utensils for people with mobility problems, and more. The exciting part of the Engineering Design and Development course is that you never know what types of projects these students will do. They become invested in the project because it was theirs from the beginning. By senior year, some of these students have decided to take a path different from how they started their journey with us, but this course still gives every student skills in research, collaboration, critical thinking, and time-management that will continue with them long after they leave.

I still remember meeting Vernell Woods, III at Open House August of 2004. My first night meeting students and families at Nansemond River High School and he told me about how much he enjoyed attending the Technology Student Association (TSA) National Convention and that he hoped that he would be able to continue with TSA now that he was in high school. His excitement was contagious, and I knew that I was right where I needed to be. I was dedicated to continuing the success of the previous advisor and let the students take the lead on competitions and community service that they would like to participate in. That year we took many students to the regional and state competitions with two students qualifying for the National Conference in Chicago, Illinois. The following years brought successful food drives, competitions, and state officer elections. Nansemond River High School was home to the 2006 Architectural Model state champions, multiple high placing individual and team competitions at the region and state level, multiple regional officers, and back to back State Presidents. Vernell Woods, III and Alec Jentink were State Presidents in 2008 and 2009, respectively. They were very active in their roles as President. Vernell was dedicated to bringing a state leadership academy for chapter officers and potential leaders. The TSA Leadership Academy is still a successful part of Virginia TSA.

Listening to students and their interests also brought *FIRST* Robotics to Nansemond River High School. For Inspiration and Recognition of Science and Technology (*FIRST*) is a competitive robotics competition that was started by inventor Dean Kamen. In the fall of 2009, I took a small group of students to the State Fairgrounds in Doswell to visit the off season *FIRST* Robotics tournament to see if they would be interested in starting a team. The students were hooked from the moment they met the team from Mills E. Godwin High School and wanted to learn everything they needed to know on how to start a team of their own. After a lot of recruiting, fundraising, grant writing, and trips to visit our mentor school- Nansemond River High School officially became *FIRST* Team 3168. The first six -week build season was a learning curve. The team had to balance their academic, social, and athletic calendars but still found their way to meetings after school and on Saturdays. Students also developed better critical thinking and collaboration skills. Not only did they have to design, build, and program one robot to play that season's game, but they also had to keep within a budget and follow the rule book that was constantly changing with updates, sometimes three times in one week. The three day competition was a world like no other. Teams filled the Siegel Center in Richmond with power tools, safety gear, robots, and energy. Following days of robot inspections, getting the hang of driving, cooperating with teams, and singing and dancing, we did not end up in the final ranking matches. Our team stuck around to watch and help other teams. In the end, we were announced during the awards ceremony as the Rookie All-Star team. This honor gave us a bid to compete at the World Championships in Atlanta, GA no matter how we placed in the rankings. Thanks to two sponsors of our mentor team, we were able to quickly secure the registration cost for the competition and only needed to fundraise for lodging and transportation with a three week turn around! Competing in the Georgia Dome was an experience that is hard to put into words. While meeting teams from all over the world, our team learned the true importance of the *FIRST* coined term-coopertition. Coopertition is their term blending cooperation and competition where value is placed on teams working together to learn from each other even though they will be both teammate and opponent at some point in the competition. This encourages students to understand that winning is not the only reason that they participate in *FIRST* and that winning is best when the team you are facing is at their best as well. The drive team was also able to meet Dean Kamen himself and pose for a picture with him. Our team brought home the Highest Rookie Seed Award from World Championships, and we had a great momentum going as we started to prepare for the next season. Nansemond River continued to compete for a few more years, but the organization changed some of its competition format and the team found it more challenging to raise funds and obtain sponsorships each year. The team put themselves on pause until they could find mentorship and sponsorship that would allow them to participate at a level that fits their goals for learning and competition.

Suffolk Public Schools brought the Sea, Air, and Land competition to our area. This competition is through Penn State, and having a local competition in our own city was an exciting opportunity for our students. TSA students that participated in this competition in Middle School brought a desire to participate while in high school. We took on the challenge, and the students looked forward to competing in the Air competition. Students practiced flying different types of drones and started to design their own grabbing tool to add to the drone. COVID put a halt to the team's progress. The team was sad but regrouped and brought together a whole new group of students excited to build and participate in the Land competition for 2023. This type of competition has been exciting for students because if the team is large they can have smaller teams with options to compete in different challenges just by choosing the environment that they would like to compete in. Nansemond River High School has offered a variety of opportunities for students to learn in the classroom and through special interest clubs, but those are not always enough to quench a student's thirst for knowledge. In the summer of 2009, Suffolk Public Schools offered a Work Based Learning internship experience for three students. During that summer, the students worked as interns with the technology company SAIC under my supervision as the district liaison. The students enjoyed their time with the company and learned a lot about how the working world functions. Sadly, we could only offer the program for one summer, but students have still found their own opportunities. One PLTW student interned with Mitre while he was in high school. Students apply for internships with NASA Langley and Jefferson Lab each year. Many students also take advantage of summer conferences and camp opportunities that are held at local universities. Many of the programs that our students participate in are brought to us by the actual university, the Virginia Space Grant Consortium, or the Virginia Department of Education. Our goal has always been to provide students with exposure to things that interest them to help foster their academic aspirations.

Student outcomes

Every student that enters my lab will be taught, no matter their intention for taking my class. Outcomes for students are more than just a grade on an assignment- it is their growth in understanding and knowledge base that helps them become who they may be in the future. We may not have the next Albert Einstein walking our halls, but we don't need him- he did what was needed from him. We have created computer science visionaries in social media app design, graphic designers (both freelance and big business employed), teachers, agriculture commodities traders, data scientists, shipbuilders, storytellers, urban planners, tattoo artists, environmental scientists, and more careers and fields that we haven't even dreamt of yet. Below are a few testimonials of student outcomes:

"Mrs. Rountree, or "Tree,, according to her students, will always be someone to whom I credit my current success. There were many instances during my academic career that I felt unimportant; Tree was and is a teacher that made her students feel safe and seen. Without Tree, I do not believe I would have pursued my design and manufacturing background. Because of her teachings, I am an Apprentice School graduate and a designer of nuclear aircraft carriers. I was lucky to be a student of hers and I am glad there are more students today that can absorb her intelligence and commitment. I am honored to have been a student of Tree and honored to know her, her husband ManTree and their shrub." **-Zachary Crabb '13** (Digital Visualization and robotics)

"I can directly trace my success in undergrad and in my Ph.D. to foundational skills and experiences I gained from my engineering classes. I think I first realized how cool physics could be during sophomore year when we used kinematics to predict the path of a compressed air rocket. In undergrad, my research advisor took me on, I think, primarily because of my experiences with *FIRST* robotics. Without that experience, I would have never worked for him or gone on to a Ph.D. I frequently use CAD skills I learned in engineering classes to build experimental apparatus. I don't think I would be where I am today without all the help, encouragement, and skills Mrs. Rountree provided me." **-Ryan Greening '15** (PLTW and robotics)

"I was able to obtain a co-op with AdvanSix, contribute to my Capstone project team being voted and awarded as the best industrial project team in our graduating class, and I am now working as a mechanical engineer on Virginia class submarines at Newport News Shipbuilding." **- Matt Clark '15** (PLTW and robotics)

"I attended Virginia Tech after high school, made closer connections with friends I met through PLTW and Robotics, and got a job using skills I learned as early as high school! My first job out of college has been at the Technology & Innovation Lab at Norfolk

Naval Shipyard, where I use Autodesk Inventor on a daily basis to 3D print solutions to problems on aircraft carriers and submarines! I'm connected to 3D printing experts in Maine, Washington, and Hawaii and have been in touch with higher-ups on establishing standards for 3D printing across the Naval Shipyards. People ask me how long I've been using Inventor, and I'm always proud to tell them it was as early as high school. I still have the wood block puzzle we made when working with tolerances and views in engineering drawings." **-Dixie Cox '16** (PLTW and robotics)

Student Testimonials

"I always looked forward to drafting classes every other day because of Mrs. Rountree's teaching style. She was hands-on, yet allowed us to be creative in our spaces in whatever way we chose. Her classes aligned well with my interest. So much so that I am now venturing into the computer design world on my own. If she was the head of an after school club, I was signing up! Mrs. Rountree made it a safe place for students to be themselves while learning and I really appreciated that! Not to mention taking the first *FIRST* Robotics Team in Nansemond River history to World Championships in Atlanta as Rookie All-Stars. GOAT status if you ask me !! **-Brent McPherson '12** (drawing and design courses)

"Dawn Rountree's strength as a teacher lies in her ability to make her classroom an open, welcoming, and safe place for students. Jock, nerd, popular, weird - typical high school labels have no meaning in her classroom. Students are all treated the same and encouraged to connect and share their skills and perspectives with each other. My time with Mrs. Rountree gave me the ability to build relationships with people different from me and to learn in a way that embraced what I was good at. I was encouraged to be me but also given the space to change and grow and I have seen her continue to do that with many students over the years." -**Camden Stevens '11** (drawing and design courses and robotics)

"After having Mrs. Rountree for 4 years and 6 classes total, the assurance she gave me about my career path, ability to explore the fields of engineering and demonstration of what it means to excel as a woman in STEM was truly inspiring. Now entering a career in software engineering I can look back on my years in her classes knowing how many skills in communication and collaboration I gained. I knew I could always count on Mrs. Rountree's advice whether personal or academic. Her class was one of the main ones in which I could truly be myself. She created an environment which challenged, but also encouraged, educational and often times humorous discussions that resonate with me even today. My friends and I still talk fondly of the memories in her classroom and you no doubt fostered such a friendly space." -**Jacquelyn Hendricks '19** (PLTW)

"I gained so much from Mrs. Rountree's classes and just from her about life in general. She helped encourage me to find my own path even if it wasn't in engineering. I was so scared to move so far out of state for school but her teachings and the mentoring I got from her helped me make such a large change to my life. I'm so blessed to have had Mrs. Rountree as a teacher and mentor." **-Christine Pinell '15** (PLTW and robotics)

"To me, Mrs. Rountree was one of my all-time favorite teachers. She showed an interest in not only the subject matter, but also in the students. She took the time to make sure that we understood the content as well as enjoyed it.

I looked forward to your class every day and the fact that it had hand-on elements like soldering circuit boards, competitions to see who can build the strongest bridge in Auto CAD, building structures out of sticks, spending a day using real surveying equipment, engaging the students by having them stand on beams to learn moment forces, and more strongly contributed to my ability to understand the concepts.

I strongly believe that Mrs. Rountree is part of the reason that I have made it to where I am in life. I think of her as a person, and her ability to connect with us strongly encouraged me to join the robotics team, which in itself has contributed to me getting my co-op in college. Furthermore, as a teacher, a person, and her teaching strategy has strongly impacted my choice to pursue mechanical engineering. PLTW and her hands on learning approach has allowed me to better understand engineering concepts that have helped me in college, as well as to help me see the "end" of my studies it college and the kind of work that I would have been doing once I finished which greatly helped me when the topics and times got especially difficult. Thank you very much, Mrs. Rountree, for being the person and teacher that you are." - **Matt Clark '15** (PLTW and robotics)

"I always looked forward to engineering classes as they were engaging and interesting compared to the usual classes offered at NRHS. I always enjoyed talking with Mrs. Rountree and I felt that she was actually invested in her students' success both academically and personally. Not to say that other teachers weren't interested in student success but it felt like she genuinely cared about us." **-Ryan Greening '15** (PLTW and robotics)

Closing

I teach with the mantra *Aspire to Inspire*. I hold steadfast to this because all interactions leave behind impressions of yourself on the other person. Like chips in a windshield, some make a larger impact than others. As a teenage girl, scared of actually going for my dream of becoming an aerospace engineer, I was inspired and empowered by my teachers and found the path that fit me the best. I would like to empower every student that passes through my life, but that is a lot to put on just one person. My desire to **inspire** students to find something within themselves has been something that always just felt unforced in my interactions and decision-making. I do not want to stop with inspiring students. Adults need uplifting support as well. This is why I love taking opportunities to present at workshops or other events. Teaching teachers is a way to help them see that the things they can do to continue to help our students grow. I have given workshops at the building, program area, district, local, state, and national levels, and I truly do it to share with others the things that they can do for and with their students. That same aspiration, with a little push from my husband, is why I joined the Suffolk Public Schools cohort of students for the Masters in Administration and Supervision with the University of Virginia, where I graduated with a 4.0 GPA in December of 2022. With the knowledge and skills that I honed while taking classes and working collaboratively with my cohort, I know that I can inspire (and maybe even empower) more adults to see their own potential, and that can trickle down to the students.

What is your aspiration?

About the Author

MRS. DAWN ROUNTREE

Dawn M. Rountree has passionately taught Technology and Engineering Education for 20 years. After graduating from Old Dominion University in 2003, Mrs. Rountree (then Zentmyer) taught at Virginia Beach Middle School with Virginia Beach City Public Schools. Following her marriage to SPS teacher Nathan Rountree, she wanted to work closer to home and found her professional home at Nansemond River High School for the last 19 years. Mrs. Rountree has been active in her school community, serving on committees, sponsoring clubs and activities, and coaching ladies tennis.

She is an active member of the Suffolk chapter of Delta Kappa Gamma, International Society for Women Educators. She is also active within her professional community and has served in many leadership positions on the Virginia Technology and Engineering Education Association Board of Directors.

Mrs. Rountree was part of the Virginia Continuity for Learning (C4L) Task Force that was formed by State Superintendent Dr. James Lane at the start of the closing of schools in March of 2020 due to the COVID-19 virus. Mrs. Rountree also holds various education awards at the district, state, and national levels.

STEM in the Classroom and Medical Field

Mrs. Sarah McDonald

I HAVE ALWAYS BELIEVED IN A HANDS-ON APPROACH TO learning. From a young age, learning science in elementary school, I would go home and perform the experiment I was taught in class. In one particular lesson, I was determined to test every chemical in my house to see if it was an acid or base solution using red cabbage as a pH indicator. Aside from making a huge mess in the kitchen and making my mom angry, I discovered that I needed to physically apply my knowledge to reinforce my learning. I determined that I learn best by physically practicing or applying concepts and technical procedures, achieving a greater understanding and gaining confidence in my knowledge. I felt empowered by true learning. There is a feeling of accomplishment and assurance through complete mastery of knowledge, and no one can take that away from you. I have a strong belief that teaching my students with this ideology of mastering the information and owning their knowledge will make them powerful and unstoppable in their path to achieving their goals. STEM education is not a trending fad in education but a pathway to activate a student's mind to achieve full understanding and encourage them to develop a mindset of self-confidence.

I was born and raised in New York City. We had the option to attend our district school or apply to a specialty program at another high school. I loved science and wanted to immerse myself in the world of applied chemistry. As a result, I chose to attend school at Thomas A Edison Vocational Technical High School in Jamaica, Queens. At Edison H.S, students were to apply for a major in the many vocational or technical programs offered; I chose medical pharmaceutical chemistry. Most of my education was hands-on and included real-life application of the knowledge I had learned. In the classroom, I had so many experiences of active learning. I remember it was a thrilling experience performing labs, like making aspirin, and watching chemical reactions and examining the factors that prevented them. I understood why a reaction happened and what would happen if it did not occur. I saw how these concepts from the labs applied to biological systems and how it could cause problems. From this I was empowered to fix these problems and understood that in learning I could help someone. It gave me the desire to search for more knowledge and wanted to do more with my education.

When I left high school, I joined the US Navy, enlisting as a Hospital Corpsman. I was eager to learn and wanted to make an impact in the world. I learned as much as I could from anyone who would teach me at the hospital. I worked in the ICU, Recovery Unit, and many clinics at the hospital. I was a sponge and absorbed knowledge. The key to my learning at the hospital was by experience, using the hands-on approach. I remembered that feeling of confidence after mastering the numerous procedures and then successfully performing them on patients. This was what made me want to learn more and do more to help my patients. After my time in the military, I started a family and began a different journey. I attended college while my family was stationed in Connecticut. I majored in biology and completed my degree. With the military came unexpected changes, and my family and I relocated to the island of Guam.

Guam is a beautiful place, where my family and I enjoyed its rich culture and wonderful people. It was the place where I grew to love my unexpected career. I wanted to enter into the fields of medicine or research, but I chose to teach high school instead. My next-door neighbor said they desperately needed science teachers and with my knowledge and experience, I should talk to someone at the local high school. After giving her suggestion some thought, I spoke with an administrator at Guam Public Schools, and they said they needed someone to teach biology and chemistry immediately. I received a provisional license and began teaching a week later. I was assigned to Southern High School in a rural island area. The high school building was in bad shape due to a typhoon damaging parts of the campus. However, the school was, at one time, a beautiful state-of-the-art facility with multiple academic and vocational learning wings, theater, multiple sports fields and a swimming pool. The repairs were slow and with island politics, the funding was never available. My classroom was on the second floor of the last academic wing, and at the end of the hall was the most unbelievable view of the ocean. The most amazing part of working at Southern High School was the students. I fondly remember walking to my classroom each morning to the sounds of guitars playing and students singing, laughing, and greeting me with "hello, Miss." I always enjoyed coming to school. In

Guam, I was immersed in learning a new culture and reaching students that did not understand what learning science could do for them. It took some effort, but I was up for the challenge, and before long I managed to develop a great rapport with my students. I wanted to empower them to use what they learned in the classroom to make lives better. I started by bringing labs into my class that brought science to life, such as the red cabbage juice indicator lab that sparked my own love of science so many years ago. I began thinking about how I wanted my students to learn and how I could get them to think about their future, community, and what they could do with the knowledge they are gaining. I wanted to be a teacher that inspires and engages students in science. I tried to add activities that would enhance the lesson while also encouraging students to think beyond the lesson. Teaching in Guam was my first experience in education, and it showed me that sharing my passion for science was what I wanted to do.

With military life comes constant change and being prepared for anything. My family was in for a great change as we moved to Virginia where I resumed my teaching career in Suffolk, VA. I began teaching 6th-grade physical science and 7th-grade life science at John Yeates Middle School. Throughout my first year, I spent time reflecting on the best types of activities to add to my lessons. I had a great mentor who was not afraid to collaborate and try new things. As a result, I learned a great deal about hands-on learning at the middle school level. During my time at John Yeates,I was fortunate to be supported by many great leaders who provided insight into what I could do and change to make the lessons better. I was very fortunate to have great science curriculum leadership that helped with ideas and supported purchasing science materials for various activities that engaged the students. I also supplemented my lessons with guest speakers that would talk to the class about different scientific professionals. I tried to promote community agencies like the Environmental Department of Quality to discuss environmental health on watersheds, companies that manage and perform environmental clean-up efforts, and even the Wavy TV 10 meteorologist. I believe it is important for students to be exposed to professionals in STEM fields, to get to ask them questions, hear their stories, and witness firsthand how science isn't just meant for the classroom but for the world. Students want to see themselves in other professionals; therefore, this exposure is impactful for them in deciding on a future educational plan and/or a future career. STEM activities and guest speaking opportunities are great for students and are often the catalyst for future decisions.

While implementing various STEM activities in my class to engage students in a new way, I was reminded that there was a great deal of planning and modification involved. Engaging STEM activities do not just happen easily like magic. There is an element of trial and error, some things work very well, and others need improvement, but this is the process of making good lessons great. The main thing is that you can think on your feet and learn quickly what needs to be modified and revised during the lesson. You must also identify what students need help with because it will be necessary to guide them through the learning process. Not all students are comfortable with having the freedom to be creative and may need encouragement in researching information. At first, students will ask you many questions, most about what they have permission to do, but then there are the serious questions, the ones that will provoke the formation of new ideas. Students will feel that they are not doing something correctly, so the teacher's role is mainly facilitation of the activity, providing reassurance that the students are on the right track. It is important to review the objectives and tasks to accomplish. STEM activities are best graded with a rubric. Having a well planned activity with listed objectives/tasks, and rubric will make the activity run more smoothly for the students and teacher alike.

Currently, I have the pleasure and honor of teaching the PLTW Biomedical Sciences Program classes, at Lakeland High School in Suffolk, VA. There are four classes in the biomedical sciences pathway: Principles of Biomedical Sciences, Human Body Systems, Medical Interventions, and Biomedical Innovations. The Biomedical Sciences Program is a rigorous academic pathway that introduces the students to genetics, anatomy, physiology, biochemistry, clinical medicine, epidemiology, and many other studies in science. Through this pathway, students learn about many career options for them to pursue as they learn the content. Students learn through projects, labs, and various activities that engage their critical thinking skills. The program teaches not only the science content but also skills that are not normally practiced in a traditional educational setting, skills that prepare students for the workforce, such as collaboration, communication, and problem-solving. STEM education is a vital teaching method for the Biomedical Sciences Program.

My teaching methods have evolved over time, but through teaching the biomedical sciences program, I've been exposed to an educational approach unlike anything I've previously experienced. My teaching is total immersion in the STEM world. I use very little direct instruction, choosing to rely on the students to research, take notes, and communicate with their peers to share information through a collaborative environment. While I do help them through difficult concepts and expand on challenging information, my students are tasked with doing the hard work of collecting data, organizing information, and communicating results. I do not have assigned seats and my expectation is that students will work with each other as there is random grouping for most assignments. I find that random grouping is most beneficial for students; they end up working with someone they would not have chosen and develop a professional respect for that individual, which is an experience that cannot be taught through direct instruction. Since I teach all four biomedical sciences courses, I have the privilege of watching the transformation of students from timid, unsure students into confident, determined academics. My methods strengthen positive attributes that drive success for the students.

Students that have adapted to this way of learning have had an overall different outlook on their education and tend to take on more challenges. I see this with my students who graduate and return to visit me, and they are eager to tell me how they were able to handle the rigors of higher-level academic courses. Even if the course is not a science course, they testify to being able to manage the difficult content and deliver on any group project with great results. The very first cohort of students has many individuals that were very successful in their undergraduate programs, and some which have moved onto post graduate education. This has been the trend with all the groups proceeding after the first cohort, students completing undergraduate programs and opting for more education with a graduate program. Even with the impact of the pandemic, though the world seems to shut down during the quarantine period in 2020, many of my students maintained their studies and commitment to their education, evident in their completion in their final capstone projects. Students that are challenged by the rigors of STEM education are more likely to extend their educational pursuits beyond undergraduate in college.

My philosophy on education is teaching students with the purpose of training them in their future career plan, must be something they love, have an interest in, or called to do (maybe leading to something greater). I want my students to leave my class wanting more for themselves because they learn that they are intelligent and have the ability. They learn by working hard and applying, even mastering knowledge, all while desiring more. Students are given this opportunity through STEM education. Their potential is realized through the process of critical thinking, reviewing/revising information, and application. This is active learning; it is not theoretical or hypothetical but realistically achieved. When students select the right answer, they know it because they see that the information is applied with success. Likewise when they are not right they see that the information is not successful. The beauty of the STEM educational approach of learning is that when a student understands that the information is not applied with success, they also see where they can change and make modifications to information and apply it differently for a different result. I believe that this is the best way to learn. I believe this because I have been nurtured in this STEM education mindset.

When reflecting on my experiences, I want my students to have the same experience and take ownership of their education and future. I believe I am teaching students that can synthesize a cure for cancer, solve environmental problems that affect our community, help patients with infectious diseases by synthesizing new antibiotics, or create a synthetic solution for patients needing vital organs. There is no goal that is unattainable when you believe in yourself and your ability. STEM education empowers students. Below I have asked a former student to share her own experiences with STEM education. Aliyyah Copeland is from the first cohort of the Biomedical Sciences Program, class of 2018.

∿

How has STEM helped me? By: Aliyyah J. Copeland

"As a 2018 graduate of PLTW: Biomedical Sciences at Lakeland High School and now a graduate student pursuing her Master's at James Madison University, I can truly attest and reflect on how much STEM has challenged, inspired, and helped me.

In high school, I had the privilege of experiencing a more rigorous and challenging course load. Of course, at the time, as a teenager, I did not see it that way. No need to sugarcoat it, but it was hard. As someone so used to grasping facts instantly and memorizing them with ease, I was frustrated at times when chemistry equations, biological concepts, and lab simulations didn't click. At the same time, though, I was having so much fun with my learning. My teachers were more so educational mentors. I use this language intentionally because they fostered learning through application, activities, and engagement in a way that didn't feel like the usual lecturing and note taking. If it had not been for those types of environments and educational mentors, I would not have the critical thinking skills and desire to dive deeper that you need to thrive in higher education.

When I came to JMU as an undergraduate Health Sciences major, I was immediately enrolled in chemistry labs, human anatomy, statistical data, and more. As you can imagine, all of those courses require intense engagement with the material, dedication, critical thinking, problem-solving, and application. In a lecture hall and lab group filled with students exposed to this for the first time, I felt confident that the textbook readings, lab methods, and lecture notes looked so familiar. I had the basic toolkit already, and it was just a matter of taking those tools with me to the store (or classroom) to find out what else I needed; while other students didn't even know what they were looking for.

Lastly, I have forever been inspired by STEM learning. At the root of every societal function and way of life is STEM. STEM allowed me to take on my education hands-on, promoted curiosity, and opened up the world to me. Just as the acronym STEM figuratively unifies science, technology, engineering, and mathematics, STEM teaches us the need to unify our strong suits in these disciplines to create, innovate, and move forward. If it had not been for STEM, I would not have felt equipped to keep going. My appreciation for STEM is rooted in so much more than the educational benefits, but also the internalized lessons of testing the limits, pushing against boundaries, and paving new pathways."

As a science teacher at the secondary level, STEM education is a goal to aspire to. I have watched many trends in education come and go, but STEM activities are a staple in the science classroom. While science can be a tedious list of concepts, we need to show the practical use of what was learned. To truly transcend the knowledge received, we must have an application piece to demonstrate that not just learning had occurred but understanding was achieved. This is where STEM fits into the Biomedical Sciences Program, a curriculum full of hands-on labs and activities to demonstrate the learning of real-world skills. Students want to see where they fit into the complicated world, possibly to help others or make positive changes, which can be seen in their choice of profession. Having the students immerse themselves in a STEM activity that challenges them to learn advanced knowledge then apply the knowledge in a lab activity allows students to see themselves in a professional role, helping others and changing the world.

My experience with STEM education at the secondary level is that many students learn more from applying their knowledge than with just memorizing the facts and figures. Students are engaged on a level that will allow them to think differently. Hands-on activities boost creativity and enhance critical thinking.

About the Author

MRS. SARAH MCDONALD

Sarah McDonald is a Navy veteran, Navy wife, proud mother, and the Lead Teacher for the PLTW Biomedical Sciences Program. She grew up in New York City and joined the military after high school. She was a Hospital Corpsman in the U.S. Navy. She married a fellow Hospital Corpsman and had two children, a daughter and a son.

Sarah attended Eastern Connecticut State University, earning a Bachelor of Sciences with a Concentration in Cell Biology. She taught at Southern High School in Guam, before coming to Suffolk Public School. She taught middle school science at John Yeates Middle School and then biology at Lakeland High School. Her current position is teaching the courses for the Biomedical Program while encouraging and inspiring her awesome students.

Section IV - The Practitioners: Central Intelligence

ONE

STEM Leadership

Dr. Stenette Byrd III

IT WAS A TYPICAL SPRING DAY AS I sat at my desk reviewing third quarter tests results when I received a call from Jessica Johnson, Director of STEM and Student Engagement (and fellow author in this book) at the Virginia Modeling, Analysis & Simulation Center (VMASC)

located in Suffolk, Virginia. She told me about an opportunity for our students at the end of the month. She shared that she had reached out to several schools but had not yet received a response. Jessica was once a teacher in Suffolk Public Schools and knew exactly what to say to get me excited about opportunities available for our students. Although numerous programs are available for students at VMASC, at the time, she was recruiting for "STEM Powerment," an Engineering Design Workshop program.

I can clearly remember all of the obstacles in the way of making this happen. It was almost time for the Virginia Standards of Learning (SOL) tests, and all of our attention was on SOL remediation. In addition to much of the budget already being spent, we would need to locate and pay bus drivers who would be willing to drive on the weekend. To make things more complicated, there was not enough time for the travel request to flow through the usual process. There were limited slots, so we would need a process to select students and alternates. And most importantly, although I was all in, I would have to convince the Superintendent that this was a worthwhile activity.

As you may have guessed, the program took place and was a big success. There was a component for parents, so my wife and I were able to attend during the first half of the program and return in the afternoon to see students complete their hands-on projects. Throughout the day, students were immersed in a virtual reality haul of a ship, and they built intricate models of the ship's piping and electricity, all while collaborating with students from other schools. To this day, I'm left wondering, how many of these opportunities are missed by timelines, procedures, funding, and other obstacles?

In Suffolk Public Schools we recognize that there are a host of intangible qualities that effective leaders must possess which are not covered in leadership courses or textbooks. These qualities are hard to explain or quantify, much like a great work of art. It is these intangible qualities that "make things happen" and prevent opportunities, like the one above, from falling by the wayside. As a result, we have started a series that we call "The Art of Inclusive Leadership."

The Art of Inclusive Leadership recognizes that there is no blueprint to create an identical copy of a great leader, nor a great work of art. This uniqueness is what makes both special. Great leaders use their own personalities, upbringings, experiences, and affiliations to inspire others. The STEM leader is no different.

A STEM-century leader should first commit to a process of self-discovery. Through this process, not only will they understand their unique strengths, talents, and limitations, but the STEM leader will also realize that some things they believed decades ago may not hold true today. It's the realization that you once believed home movie night would involve a car trip to the video store that allows the leader to open their eyes to the fact that we do not have the capacity to know for sure what movie night will look like two decades from now. STEM education is the vehicle to prepare students for a world that we cannot predict and is synonymous with innovation.

Through our "The Art of Inclusive Leadership" series, Suffolk Public School leaders participate in a self-identification protocol that helps them understand the strengths and limitations of various personality types. By understanding their individual value and the value of others, all viewpoints and contributions can truly be appreciated. Through this series, the leader also develops a confidence that, although an entire room of people may view a situation differently, their thoughts, feelings, and contributions are valuable to any discussion. And more importantly, discussions are more powerful when an inclusiveness of ideas are also present along with race, gender, and other forms of diversity. This understanding makes others feel safe to respectfully challenge the thoughts or actions of superiors, policies and/or practices. I remember reading a story of an airline co-pilot whose cultural beliefs prohibited him from challenging the authority of the higher-ranking pilot. The co-pilot's deference to the pilot would not allow him to question further the decision to stay the course, and the result was catastrophic. Although this is an extreme example, everyone in a school building should have a voice and feel that their contributions matter.

In addition to the understanding of one's identity, inclusive leaders understand that learning and opportunity gaps happen naturally unless we are intentional about efforts. For example, I live close to the VMASC and could have easily driven my twins to the STEM Powerment events. However, by providing transportation, the event was open to a more diverse group who may not have been able to provide transportation. Not only did we provide an opportunity for a group of students to experience virtual reality for the first time, consider the added value of the conversations and experiences of those whose life circumstances are different from my twins.

Policies, regulations, and rules are designed to create order, ensure safety, and move people in a certain direction, and it is the leader's job to ensure all are followed. However, the leader must also acknowledge that policies, regulations, and rules may be barriers to innovation which is at the heart of STEM education and the heart of, what we call in education, the 5 C's (Critical thinking, Creativity, Communication, Collaboration, and Citizenship).

The STEM-century leader must often walk a fine line as policies and procedures play catch-up to innovation. Sometimes our only sense of comfort is that we make decisions based on what is best for the child. STEM opportunities come in many forms and may bump heads with conventional education. Imagine your students have the opportunity to collaborate with a leading engineer as they begin the design process for building a new school in your division. The architectural firm is 32 minutes away, on the other side of the state's border, and students are requested to bring their laptops to the collaborative meeting. How does the leader respond when policies deter trips over 30 miles, buses are needed for out-of-state trips, and firewalls prevent programs from being downloaded on computers?

The STEM leader's job is to remove barriers through courageous decision-making and giving other leaders permission to take risks and, at the very least, ask questions. In addition to this aspect of leadership, community partnership is at the heart of STEM education. Although these partnerships will be covered in a separate chapter, the leader must have the tools and skill set to cultivate and nourish these partnerships. These skills involve visibility within the community, authority to make decisions, and an attitude of persistence.

Let's take a moment to reflect on the inclusive leadership traits that were needed to ensure the opportunity in the opening paragraph was fulfilled. These are the same traits that necessary to promote STEM education in your school or division:

- Visionary Leadership
- Courageous Leadership
- Political Savviness and Relationship Building

First, the leader must have a vision for STEM education. Visionary leadership is a process that ensures other's buy-in to and support your vision. Buy-in cannot be achieved by giving a directive. Buy-in comes from others understanding the "Why" behind your vision and truly believing that your vision is in everyone's best interest. Once achieved, the leader has soldiers who are ready to spread the vision, look for opportunities to implement, quiet the naysayer, and smash through barriers.

Courageous leadership is twofold: 1) Do you have the courage to break barriers and continue the fight when others may not see your vision? 2) Do those who work for you have the courage to push on you when they feel that your eyes may be closed to an opportunity? Those who lead by fear may have highly compliant organizations; however, as stated above, compliance can be a barrier to innovation and harm STEM education.

The last trait of the STEM leader is political savviness. Early in my educational career, I remember saying, "I'm not into politics; I just want to do what is best for my students." Although this may seem admirable, having political savviness is the skill needed to bring big opportunities to your schools. The people that I have met and the relationships that have been formed by attending community events and other highly political events cannot be underestimated. I encourage you to look at the list of sponsors and contributors of this book. This could not have been achieved by sitting in our office. Many of the relationships between Suffolk Public Schools and the contributors and sponsors of this project evolved at community events, civic involvement, board affiliations, alumni events, fraternity/sorority involvement, and one connection even dates back to 1993 when I first met my wife, but that is a much longer story for another time!

I encourage you to sit with your leadership team and ask the following questions:

- Have I gained the political savvy within my community to bring STEM opportunities to my school?
- Do those who report to me have permission to bring opportunities to me for consideration?

- Do we have policies that discourage STEM and other opportunities for our students?
- How have we promoted/publicized the work in STEM that is currently taking place in an effort to encourage more opportunities?

As you reflect on the questions above, keep in mind that political savviness is a skill that is developed over time, as you experience previous conversations and decisions that you have made, morph into actions that are totally unexpected. The politically savvy leader has learned to manipulate and predict these outcomes. My advice is to spend time observing those leaders with this skill set and begin asking probing questions.

In closing, look for barriers, real and perceived, that could stifle conversations that may lead to opportunities for students. In an ideal situation, teachers promote the great work of their students, not only on social media, but also in other social situations. This growing hype results in your students being asked to perform at an event in which the director is confident that he/she has your approval. At these events, conversations take place that produce partnerships that result in an economic opportunity for your students.

About the Author

DR. STENETTE BYRD III

Dr. Stenette Byrd serves the Suffolk Public School Community as Chief of Schools and is responsible for supervising the Department of School Leadership and Innovation, including supervising the division's schools and school leaders. The Department of School Leadership and Innovation encompasses the departments of Elementary School Leadership, Secondary School Leadership, Career and Technical Education, and Technology.

In this role, Dr. Byrd performs complex professional and administrative work and assists with the supervision of division-wide activities with emphasis on principal supervision and leadership, informational technology, and career and technical education. He supervises directors, coordinators, principals, and other school and division administrators for numerous programs.

In addition to his role as Chief of Schools, Dr. Byrd is an adjunct professor at Old Dominion University and co-facilitates a Professional Development Network composed of Region 2 Chiefs and other division leaders, who are tasked with principal supervision, extending the superintendent's vision, developing future leaders, and improving climate and culture within schools.

Prior to his current role, Dr. Byrd held the positions of Executive Director of Elementary Schools (Newport News, VA), Director of Secondary Schools (Suffolk, VA), Elementary School Principal (Isle of Wight, VA), Middle School Principal (Isle of Wight, VA), High School Principal (Isle of Wight and Suffolk, VA).

Byrd Byrd is married to Tonya Byrd, and they are the proud parents of twins, Stenette Byrd IV and Trae Denise Byrd.

TWO

STEM Resources and Supplies

Mrs. Wendy Forsman

IN ORDER TO DISCUSS STEM RESOURCES AND SUPPLIES, ONE must first understand the nature of the relationship between Finance, Purchasing, andInstruction. Finance and Purchasing departments exist to support Instruction. For non-instructional people to fully understand the needs, resources, and supplies of STEM instruction, they have to be at the table from the beginning of the planning process. Then, equipment needed can be bid, orders placed, allocations adjusted, grants sought, and resources dedicated. All of this is dependent upon these two departments collaborating well.

School division Finance and Purchasing is often viewed by Instruction as inflexible and as operating on the basis of "play by the rules," as compliance is our focus. However, providing for the needs of STEM-based instruction requires flexibility to plan and execute lessons. There are multiple layers of considerations for the Finance and Purchasing departments such as: a realistic need for on-time resources, centralized or decentralized purchasing, contracts with vendors, the existence of a warehouse, purchase cards versus encumbrance, and allocation systems in place. Centralized purchasing involves gathering what is needed and having the Purchasing department deliver one purchase to multiple locations. Decentralized purchasing involves each location making its own purchase individually. Each way has costs and benefits that need to be evaluated before deciding what works best.

Realistic need for on-time resources

STEM instruction requires more than the traditional resources and as such, more planning and collaboration must be given to the needs of Instruction. Recurring costs of consumable items may be allocated to schools based on the number of students, whereas reusable supplies may be centrally budgeted. Detailed lists should be generated and vetted by instruction to ensure every teacher and student has the supplies needed to support STEM instruction fully. The age of the building is another consideration due to the availability of collaborative space, internet access, and use of electricity and water, such as sinks in laboratories. The Facilities study that was completed in 2021 showed the division areas where programming did not match the needs of Instruction. Smaller classrooms could not accommodate the project learning desks, computer labs with desktop computers were outdated, and older media centers did not have collaborative spaces. The addition of our Center for Performing and Production Arts (CPPA) required architectural design to allow for the removal of lockers that became collaborative spaces with benches and ways for students to sit in small groups and collaborate. Special sound booths need design and permits from the city so that electricity and HVAC equipment can be run to them. Special lighting and paint are necessary to create a black box production room. These examples show that special programs may need design work from an architect, city approval of planning and design, and construction. Older computer labs or media centers may need to be reconfigured into collaborative spaces with different furniture and fixtures that are flexible in use and nature. Nested tables with lockable wheels, stacking chairs, project learning desks, and soft seating may require more resources and planning that could cross fiscal years. Supply chain issues must also be considered in timelines. Some furniture and fixtures can take up to one year to be delivered.

Allocations

The division has specifically set aside funding in the budget for allocations to each school based on student count. Our division spreads the allocation out by transferring 80 percent of the money in August and the remaining 0 percent in January. This approach helps schools pace their spending. These funds are for instructional supplies in the classroom. The total budget is multiplied by 80 percent and then divided by the total enrollment from the previous June. This gives us a per-pupil amount to fund in August. The amount per student is then multiplied by the total enrollment by school and the total is transferred to the school. This is repeated in January with the remaining 20 percent using the September enrollment.

In addition, specific funds have been budgeted for science, musical instruments, and technology. These funds are spent centrally to maximize the purchasing power by obtaining quotes. Each year, the science and technology department heads evaluate needs and request budget allocation changes based on those needs assessments. All other allocations are assessed on inflation.

Centralized or decentralized purchasing

Having teachers list all their needs before the semester begins and then purchasing those items centrally offers considerable cost savings. However, needs change and new opportunities mean that the STEM instructors can end up without timely supplies under this method. Allowing each instructor to purchase as needed is very costly. A hybrid system of gathering the data on supplies for specific planned lessons and a purchase card system that allows on-demand purchasing gives instruction the flexibility needed to provide all resources. For example, an instructor working on a lesson involving perishable food would not be able to have it purchased once a semester and would need access to a purchase card to get the supplies "just in time" for the lesson. Lessons that use metal plates, legos, or even preserved lab samples can be purchased well in advance. A well-controlled purchase card system allows our schools to buy supplies and reimburse the district for the purchases from their allocated resources.

Contracts with vendors

Contracts with specific STEM supply vendors negotiated to allow for one-day delivery or free delivery of goods can save the district money and give the instructional staff the ability to control when and how needed items are delivered, saving storage space at the school classroom level. Delivering general items once a semester, such as gloves, aprons, and multi-use items saves the division money and allows Instruction to focus on items individual to each lesson. Divisions with warehouse space can keep inventories of supplies and drop them at schools on a regular basis. School divisions without warehouses have to rely on vendor contracts and purchase card systems to get the supplies delivered "just in time" for the instruction to take place.

Encumbrances versus purchase cards

Traditionally, encumbrances require someone to type in each item, vendor name, amounts, and quantity, have been the only way for schools to get supplies needed for instruction. Encumbrances start with a requisition that moves through an authorization process and becomes a purchase order that encumbers or saves a portion of the budget to pay for goods and services. The introduction of google sheets (like an Excel spreadsheet) where all teachers can collaborate online and enter all items needed have brought purchasing into the 21st century. These sheets can be opened, all entries made by the teachers, and the sheet can be closed or locked by a deadline. The items are then sorted, and quotes from various vendors can be sought via email. The lowest quotes for each item can then be awarded so that multiple vendors can supply the goods at the lowest cost. Twice annually, this process is repeated for items that are multi-use or for required labs. The Purchasing department uses a combination of purchase card and encumbrance to fulfill these orders. Schools regularly have need for items less than one week in advance for the next week's STEM opportunity.. The instructor identifies what is needed and the bookkeeper requests use of the purchasing card to get the items shipped directly to the division.

Example: Centralized

Planned Activities:	# Students	Cost per unit	Vendor	Total Cost
STEM Lab Kits Gr. 6-8	3,500	$5.36	ZB Vendor	$18,760.00
Basswood Bridge Building kits Gr. 6-8	1,113	$7.42	JL Vendor	$8,258.46
Weather instruments K-5	4,625	$6.10	XYZ Vendor	$28,212.50
Simple Machines Gr. 3-5	3,250	$10.22	BL Vendor	$33,540.00

Notice that the instructional staff knows which grade levels, how many students, what supplies are needed for each activity, cost per student to have the lab or activity, vendor (even though Purchasing will get quotes and choose the lowest one), and total cost. These items are purchased in bulk after quotes are received. This is centralized purchasing.

Example: Decentralized

Vendor	Description	School	Amount
AM Mktp	CTE Materials STEM Lab S. McDonald	NRHS	$124.95
SS Vendor	CPPA Materials special lab	LHS	$244.88
WM Vendor	Class supplies	BTW	$264.65

Summary

STEM planning can be challenging if Instruction and Finance do not work together to break down barriers that delay the delivery of needed materials. Working together to understand timelines, types of materials needed, and what items are perishable can make mixed-use delivery possible. A clear understanding of the programmatic needs from the initial design through the delivery of instruction helps make seeking grants, identifying allocation funding, and procurement in a timely manner possible. In short, understanding, collaboration, and planning are key to providing STEM resources and supplies.

About the Author

MRS. WENDY FORSMAN

Wendy Forsman currently serves as the Chief Financial Officers for Suffolk Public Schools in Suffolk, Virginia. She has 24 years of experience in public school finance. She is responsible for the areas of Finance/Purchasing, Food and Nutritional Services, Facilities and Maintenance, and Wellness/Benefits. She is a graduate of Virginia Wesleyan College and earned her Certified Public Accountants license in 2004.

THREE

STEM Infrastructure

Terry Napier

WHEN CONSIDERING THE IMPLICATIONS THAT STEM-RELATED programming has on both building design in new construction projects and repurposing and renovation of existing spaces in older buildings, it's important first to understand what STEM will look like in your school or school division. How new or existing buildings and spaces need to function to ensure the building or classroom layout and features are an asset, rather than a hindrance, in meeting the overall program objectives is essential to the success of any STEM program.

Let's begin by assuming that a STEM curriculum has already been developed in your school or school division. Let's also assume that students will be involved in both individual and collaborative lessons and projects that require them to engage in problem-solving, critical analysis, teamwork, creative thinking, independent thinking, verbal and written communication, and all manner of hands-on and project-based interactions that integrate science, engineering, math, and technology instruction.

Given the parameters just outlined, let's now look at 11 considerations related to building and classroom design layouts that would enhance the effectiveness of the STEM curriculum.

1. Increased Space

Traditional classrooms generally range anywhere from 800 to 1000 square feet in size. Given that STEM instruction involves collaborative group work and active interactions, more space is better. While this may not be possible with existing buildings it can certainly be accomplished in new buildings by designing large and well-planned collaborative areas.

2. Flexible Layout

Demonstrations, projects, and team-building activities will often be the order of the day in STEM classrooms. Classroom spaces with adequate size, flexible seating and furniture, and cabinetry designed to maximize open floor space allow for the instructional area to be rearranged with ease and quickness to accommodate the day's activities, whatever they may be.

3. Material, Supplies, and Equipment Storage

Adequate and organized storage areas are essential for STEM classrooms, given the quantity, type, and size of materials, supplies, and equipment required for STEM instruction. Storage rooms or closets separate from the instructional space are preferable whenever possible and should be large enough to make access to all materials, supplies, and equipment as easy as possible. In existing classrooms, adequately sized storage cabinets designed to provide easy access to their contents and anchored along perimeter walls to maximize open floor area would be beneficial.

4. Project Storage

STEM projects cannot always be completed in one or two instructional sessions. Consideration needs to be given to where ongoing projects can be stored so they do not interfere with other class groups or impede on floor space required for daily lessons and activities. As discussed earlier, the ideal solution would be a storage space separate from the instructional space. Again, this could easily be accommodated in new construction but could pose a problem for existing buildings and classrooms. In situations where rooms are being repurposed, the design and layout of cabinetry could accommodate project storage, or possibly an adjacent classroom could be converted to a storage area and utilized by several classrooms.

5. Large Collaborative Learning Areas

Having adequate space in the STEM classroom and the ability to "create" a large group instructional space when needed is essential whenever teachers need to gather all of their students into a more traditional arrangement for introduction to concepts, providing project instructions, or having whole group interactions and discussions. This can be accomplished by design or rearrangement of furnishings, but providing the appropriate space and furnishings to quickly accomplish this classroom transformation is essential.

6. Verification Areas

STEM projects often involve robotics, chemistry, electronics, and physics. Once an experimental type project is nearing completion, students need an area where they can do some live trials of whatever the project is intended to perform. This area needs to be situated in a location that does not interfere with or encumber the ongoing work of other students in the classroom.

7. Printer Stations

Many STEM activities require students to print materials, specifications, design ideas, etc. To facilitate this process there needs to be a specific area designed into the overall classroom scheme where multiple printers and printing supplies will be maintained. This allows students access to a very basic function in a specified area and keeps that process from interfering with the other workspaces in the classroom.

8. Presentation Space

Teachers know how proud students can be of their accomplishments, and they like to share their knowledge and what they've learned. A specified "presentation" area in the classroom that is also anchored to an exterior wall allows students to engage in both small and large group presentations by simply rearranging the room quickly to accommodate either whole class presentations or smaller workgroup types of presentations by simply moving around tables and chairs.

9. Project Display Areas

Like most students, STEM students like to see their work on display. They are proud of their projects and want others to see the fruits of their labor. Fixed display cabinets provide an opportunity to place student work front and center for all to see. These areas can be designed into wall structures or mounted as surface installations, but it needs to be accomplished in a way that does not infringe on the work or instructional areas.

10. STEM Classroom Furniture

Furniture in a STEM classroom must be flexible and easily rearranged even during an ongoing lesson. Tables, chairs, carts, and storage cabinets should have casters for easy movement. Chair seating materials and work table surfaces should be durable, easily cleaned, and light in color to ensure spilled materials, chemicals, small project parts, etc. are easily visible. Tables should be shaped to allow multiple configurations that accommodate individual, large-group, and small-group configurations. Seating height on chairs, tables, and stools should be varied and/or adjustable to accommodate student needs and preferences depending on the tasks.

11. Building System Considerations

Electrical Service

It would be rare to find a traditionally designed school building with adequate electrical service components to service a STEM classroom properly. STEM lessons and projects involve many different types of electrically powered equipment. This requires not only adequate electrical outlets but also adequate voltage and amperage serving the STEM classroom. This would include additional 120-volt circuits to power the numerous electrical outlets that will be required. Outlets need to be installed in both the floor and the walls whenever possible. The placement and number of electrical outlets should not preclude the STEM space from being as flexible as possible. 220-volt service is not something that is usually found in traditional classrooms but if the STEM program will require any 220-volt equipment the room needs to be equipped to accommodate that need.

Classroom Lighting

Classroom lighting is important to occupant productivity and safety. STEM classrooms should be equipped with LED fixtures and bulbs that provide whole-space lighting and also provide task lighting in specific areas when needed with separate controls for each area and fixture type. Project areas may require more intense lighting at times than other areas of the room. Ideally, proper lighting can be maintained in multiple areas at the same time and not impact surrounding areas. Dimmer controls on all fixtures would allow even more flexibility in classroom lighting.

Plumbing Fixtures

Additional sinks with industrial-grade faucets and spray heads installed in various locations near the proposed project areas would be beneficial for both clean up and meeting water needs that various collaborative groups will have during project work. Traditional primary classrooms usually have one classroom sink to meet all needs. This is insufficient and becomes even more of a problem in a STEM classroom.

HVAC Equipment

STEM classes would benefit from higher volumes of conditioned air being introduced into the room, along with an increase in outdoor air cycling. This is especially true during cooling cycles. Utilizing an increased number of electrically powered devices coupled with computer equipment creates a higher heat load on any space. The additional cooling capability would not only keep the classroom more comfortable during active collaborative work sessions but also provide a cooler environment for the equipment being used.

Classroom Finish Materials

Wall, floor, counter, cabinet, and ceiling materials need to be upgraded to the most durable and cleanable products possible. An active group of STEM students would be expected to have frequent spills and messes while engaging in individual and group projects. Utilizing materials that are required in many science-related projects can be difficult to clean on traditional wall, floor, and ceiling surfaces. This becomes more exaggerated when there are multiple projects being worked on at the same time. It's important that STEM workspaces be as cleanable as they are flexible and that any cleaning operation required be completed quickly to ensure no time is lost when transitioning between student groups.

Finally, as schools and school divisions place greater emphasis on STEM programming and instruction, it becomes essential to design, build, renovate, and repurpose classroom spaces in a manner that enhances the STEM efforts undertaken. We need to ensure that our design and renovation efforts provide classroom spaces that allow creative, engaging, and rigorous collegial interactions among students and teachers. It is also important to recognize there is no concrete blueprint or template when considering what a STEM classroom might look like or how it may function in differing schools and school divisions.

The purpose of this chapter was to provide the reader with discussion bullets and points to ponder when considering the impact of building design or layout when planning a STEM program and curriculum. All of us that have devoted our careers to education have long ago realized that one size certainly does not fit all and that holds true for STEM programs the same way it does for all other programming associated with the educational process, whether it be instructional program development or infrastructure design considerations.

About the Author

TERRY NAPIER

Terry Napier currently serves as the Director of Facilities and Planning for Suffolk Public Schools in Suffolk, VA. Terry has earned a Bachelor's Degree in Special Education, a Master's Degree in Educational Administration, and certifications as a Facilities Management Professional (FMA) and Sustainable Facilities Professional (SFP) through the International Facility Management Association (IFMA).

In addition to his current position, he has also served as a Special Education Classroom Teacher, Elementary Assistant Principal, Elementary Principal, and Assistant Director of Facilities and Planning, all with Suffolk Public Schools.

Terry has worked in public education for 33 years.

FOUR

The STEM Network

John W. Littlefield

"Difficult to see, Always in motion is the future."
—Yoda, The Empire Strikes Back

WHEN I, JOHN LITTLEFIELD, DIRECTOR OF TECHNOLOGY, arrived in Suffolk Public Schools 25 years ago only a few computers existed in the district. Some were in the libraries and some were in administration, but none were connected to the internet. Suffolk was significantly behind back then because of the lack of technology personnel and funding. Joyce Trump, Superintendent, and Dr. Liverman, Assistant Superintendent, believed that technology did have a place in education and that a plan was needed to move forward. I was hired to develop a plan and move the district forward. Over the course of three months, I reviewed the needs of the school division and created a plan that would provide a foundation that the district could sustain. We then were able to hire staff, create a technology budget, and over many years, build a fully-staffed Technology Department.

The strategy was to build an up-to-date, fast, and reliable technology infrastructure first, and then we could layer on top of the resources needed to move Suffolk Public Schools forward. First, we must understand what "the network" is. In the Technology field "network infrastructure" is described as the interconnection of computers, machines, and operations. In order to provide this "interconnection" Suffolk Public Schools started with the idea that we need something fast and reliable. The initial idea was to' lease network connections between the schools; however this was very expensive, so we looked for other options. Later I will describe the solution we are currently utilizing. The foundation of the SPS network is a fiber infrastructure that interconnects all schools with adequate bandwidth capacity to allow for growth and capacity for the instructional needs of the division.

Today our fiber network consists of private fiber and iNet fiber (Spectrum/Charter dark fiber). Early in the planning, Suffolk Public Schools was able to leverage the settlement agreement between Falcon Cable and the City of Suffolk that made provisions for iNet fiber connections to public buildings. The agreement required the cable provider to provide four strands of fiber to each building, with the only cost being the "last mile". "Last mile" is the term used to describe the distance or path between the public backbone and the actual building needing the connections. At some locations, this may have been a few hundred feet or several miles. Suffolk Public Schools spent approximately $450K (over several years) building these connections to the schools with Charter/Spectrum. We started with the schools that were easily connected and the backbone fiber was readily available from Charter/Spectrum. The schools in the rural parts of Suffolk were completed last since Charter/Spectrum did not have fiber resources available in the Whaleyville and Southwestern areas of the city (these were also the most expensive connections). We are now building private fiber pathways that will replace the iNet fiber with private fibers that are owned by Suffolk Public Schools or Suffolk City. Currently, we have private fiber pathways to the following schools: Northern Shores Elementary, Colonel Fred Cherry Middle, Creekside Elementary, Florence Bowser Elementary, John Yeates Middle, Nansemond River High, Mack Benn Elementary, Elephant's Fork Elementary, Hillpoint Elementary, King's Fork High, King's Fork Middle, John F. Kennedy Middle, and Southwestern Elementary. We are working with the City of Suffolk to continue to build the remaining pathways to our other schools. This will give us redundancy for the connections to the schools and reduce our dependence on the Charter/Spectrum fiber. This fiber network uses the three high schools as hub sites to distribute connections to the nearby elementary and middle schools. This Wide Area Network (WAN) has enabled Suffolk Public Schools to achieve opportunities that would have otherwise been cost prohibitive. Currently, Suffolk Public Schools does not have monthly recurring costs for the WAN interconnecting the schools and support buildings which enables us to utilize these funds for other things. This is a significant cost savings for the school district annually.

On top of the fiber Wide Area Network (WAN) are the Local Area Networks (LAN), consisting of wired and wireless networks in each building. The wired network provides secure connections to desktop computers, specialty systems, security cameras, access control, wireless access points, and other building control systems. The wireless network provides secure connections to over 16,000 staff and student wireless devices. The wireless network has been meticulously designed by our network team to provide coverage in every classroom, other educational spaces, and offices. The entire network is monitored 24/7 to ensure that outages are minimal. With the backbone of the network established, Suffolk Public Schools is able to focus on providing educational resources that are able to be delivered with ease.

To ensure the user experience is fast and reliable, each computer system and application is evaluated before purchase. The evaluation process for hardware includes durability, performance, longevity, and cost. Additionally, each software application is reviewed for compatibility with the SPS network environment, dependencies, compatibility with SPS user devices, has appropriate support, and is instructionally-relevant. These evaluation processes have proven to save costs and ensure that Suffolk Public Schools is providing resources that are reliable and appropriate for the instructional needs of our students.

The connection to the internet is mission-critical for K12 education. Suffolk Public Schools has seen significant increases to the bandwidth of the internet connection over the past 25 years. In the early 2000s, we had an initial internet bandwidth of 56K, and today we are utilizing 5Gb with plans for the 2023 -2024 school year to increase to 10Gb. Having a fiber connecting all the schools is why Suffolk can purchase and manage internet capacity centrally. This eliminates the need for maintaining multiple sites internet connectivity while maintaining quality service to each school and classroom. The bandwidth is monitored constantly to ensure there is enough capacity for the ever-increasing needs of the classroom. Currently, Suffolk Public Schools utilizes eRate funding to reduce the monthly recurring cost of internet access.

We also host approximately 125 servers that host instructional and business applications for our staff and students. This server environment is built on physical servers and blade servers using a virtual server environment that provides redundancy and capacity. These servers are housed in the data center located in our Technology Building on the John F. Kennedy Middle School campus. Over the years we have added a backup generator, additional cooling capacity, and power conditioning to ensure a fast and reliable data center. In addition to the servers that are hosted locally, Suffolk Public Schools is utilizing some cloud hosting services such as Amazon Web Services for multiple applications and for some system backups. We also have chosen to use a hosting service for our district website, to prevent interruptions of service due to local issues related to weather or other natural events.

Interactive Whiteboards with a Chromebox are available in every classroom to ensure each teacher has state-of-the-art technology in the classroom. These systems make use of short-throw laser projectors to provide the largest image size possible for the students. The classroom system also includes audio and wireless keyboard/mouse to provide flexible resources for the teacher. The Interactive Whiteboard systems provide multi-touch interaction for the teachers and students. Additionally, each school has video conferencing systems and document cameras available that can easily be connected to the classroom system to bring the world to the classroom.

Suffolk Public Schools has chosen to utilize the Google Suite for Education for both staff and students. This resource provides an efficient and effective productivity tool. The use of Google Gmail, Drive, Calendar, and other resources prepares our students with 21st-century skills to make them successful in today's online world. Google is not the only resource available as Adobe, Microsoft Office, AutoCad, and many other resources are available to the staff and students.

All students have access to Chromebooks to use in school or at home. Four years ago Dr. John B. Gordon III, Superintendent, moved the Technology pointer significantly by enabling one-to-one Chromebooks for all students K-12. Then the COVID-19 pandemic came, and it was all about being in the right place at the right time and Suffolk Public Schools was ready for the shift in learning. The only challenge that SPS had was the lack of home internet access which we were able to mitigate with mobile hotspots from Kajeet and T-Mobile (Project 10 Million). We created a checkout process for the mobile Hotspots that ensured the students who needed it could access the resources. Suffolk Public Schools was able to continue the learning for all students in spite of the challenges that COVID-19 brought.

To ensure the thousands of devices and infrastructure are operational each day, there is a team of technicians who are in the background maintaining the technology. There are eight hardware technicians who are at the bench repairing hundreds of Chromebooks each week and two network technicians who are in the field daily maintaining the hundreds of routers, switches, wireless access points, and other IP-connected devices throughout the district. We also have four Information Systems Technicians that provide HelpDesk support, student information systems support, and prepare local/state reports. To round off this team, we have two Server Specialists and two Application Specialists supporting the hundreds of servers and applications. To provide oversight of staff there are two Coordinators, one oversees the Information System Technicians and Instructional Technology Resource Teachers (ITRTs), and one oversees the

Hardware & Network Technicians and Server & Application Specialists. This highly skilled team holds a variety of Technical Certifications to ensure they are well prepared for the daily challenges.

We also have to ensure that all our teachers, students, and support staff can utilize the Technology resources that have been made available. Suffolk Public Schools has seven Instructional Technology Resource Teachers (ITRTs) that provide face-to-face, virtual, video, and other resources to enable the use of Technology. These seven ITRTs are a vital part of the Technology Department and are in the schools daily working with staff and students. They are former classroom teachers that are Instructional Technology Integration experts with a variety of different teaching backgrounds. As a team, they are able to support teachers "where they are" and can easily relate to their needs. They have created an arsenal of creative resources that include: Techno While You Go, PopUp TechTalks, Tech Bytes, Technology Coaching Menu, Edu badging, and an active social media presence. These resources are constantly being updated on their website.

The COVID-19 pandemic also pushed Instructional Technology to new horizons. With Zoom, Google Meet, and BigBlueButton, video conferencing has become commonplace in instruction. Additionally, the Virginia Department of Education has provided Canvas, a digital learning management system to every school district in Virginia. Suffolk Public Schools was positioned well to roll out these web-based systems to our students and staff. We have seen significantly increased bandwidth demand. Fortunately, we were prepared for this additional bandwidth demand and are just now seeing the need to increase the connection to the internet. Internally our network has adequate bandwidth capacity to sustain this and future growth. The Application Specialists were able to quickly build the integration for these new applications and the ITRTs were ready to train our staff to utilize these new resources.

Digital Signage Technology is helping Suffolk Public Schools share information with our staff, students, and visitors. Last year we installed over 100 digital signage displays in the main entrances, lobbies, and cafeterias of all schools. These displays are managed by RiseVision, a digital signage software that provides over 500 K12 educational templates that can be customized by each school. They can be scheduled by school staff to deliver timely content during the school day based on what may be going on at any given time. Content can also be pushed out by the district Community Engagement Department to all displays or to a particular school if a special notification is needed. Students find the content interesting and engaging as they are moving throughout the school.

In summary, Suffolk Public Schools has a solid technology foundation that is based on strong network infrastructure and skilled technical staff. Each year we meticulously plan for the replacement and additions of network-connected equipment and applications. As new technologies emerge, we will continue to look forward to the best solutions that will sustain the needs of instruction. The Technology Department has received positive feedback from the teachers, students, and parents that are surveyed each year with regards to the instructional technology that is available. The entire team in the Suffolk Public Schools Technology Department is dedicated to providing fast and reliable technology resources that support student learning. If we continue to plan out five years every year, we will continue to be prepared for the future. I can only imagine what the future holds and how the network will look in the next 25 years.

About the Author

JOHN W. LITTLEFIELD

John W. Littlefield, Director of Technology, joined Suffolk Public Schools in January 1998 to move the Technology needle for the district.

Prior to Suffolk Public Schools, John was with Isle of Wight County Schools, overseeing their Technology support and modernization. While at Isle of Wight County Schools, he was able to help them roll out local & wide area networks and video distribution systems at all of their schools.

Prior to working in public education, he worked in the retail market supporting, installing, and selling computer systems. In the early 1980s, he also worked as an electronics technician for a contractor for Wallops NASA facility on the Eastern Shore of Virginia.

John has an Associate's Degree in Applied Science with a Major in Engineering/Industrial Technology. Additionally, he has earned a variety of Certifications in Technology and Project Management.

John currently lives in Windsor Virginia with his wife. When he is not working you can find him enjoying his family and working in his yard.

FIVE

STEM Careers

Jessica Avery

I have worked in Suffolk Public Schools for nineteen years. Like many other teachers, I did not begin my career in education through a typical college teaching program. Prior to beginning a career in education, I was in the United States Navy. While in the Navy, I received a bachelor's degree in criminal justice. I always believed I would work in the criminal justice field. However, after leaving the Navy, I decided to become a special education teacher. I took the required special education courses that allowed me to obtain a provisional license. I was hired as a high school special education teacher. After the first year, I realized I really enjoyed working with students and decided to go back to school for a masters in elementary education. After obtaining my certification in elementary education I began teaching fourth grade, and I absolutely loved it! A few years passed, and I got the itch to lead and added the administration and supervision endorsement to my license. On July 1, 2021 I moved to the Human Resources Department. These different positions have prepared me for the understanding of hiring staff, what qualities to look for, and the ability to recognize those who are coachable.

A large component of my department's responsibility is recruiting, onboarding, and retaining staff. Recruiting and retaining in particular can be challenging; however, we continue to think about ways to improve in these areas. Onboarding is an area where my department works hard to connect with new staff. This is our opportunity to welcome new employees to our division and to make them feel welcomed. When they leave our department, we want staff to know we are excited that they are joining our team. Prior to recruiting, we need to understand what positions we are hiring for and what the qualifications are to fulfill these positions. In November 2022, a report to the Governor and the General Assembly of Virginia found recruiting and retaining qualified teachers after the COVID-19 pandemic has made it harder for divisions to fill vacant teaching positions. Leadership from school divisions in Virginia were surveyed and 94% percent said it has become harder to recruit classroom teachers, and 90% indicated it has become harder to retain teachers. In Virginia, the teacher workforce are leaving and fewer are becoming licensed for the first time. During the initial part of the pandemic teachers leaving the profession declined. However, it began to rise during the 2021-2022 school year. Teachers leaving the profession was 12% higher than prior to the pandemic. In addition, teachers getting licensed for the first time declined by 15% lower than prior to the pandemic.

The Human Resources Department is responsible for recruiting and screening applicants to ensure they are qualified, sharing information with school administrators to ensure they are privy to applicant information, making offers, and onboarding all staff in the division. Specifically, we look at the licensure of all our teacher applicants. For this reason, we screen applicants for principals to ensure applicants qualify for a teaching license. Unfortunately, because we do not have a large number of applicants seeking to teach math and science, finding eligible applicants to teach these courses can be difficult. What does this mean for school divisions? Human Resources Departments now have to find ways to staff teaching positions in an environment that data shows is more difficult than it once was. One avenue to help applicants that are interested in teaching math or science is sharing information about the praxis tests. If they take and pass that they will be eligible for a provisional license.

This year we started the year off with many open positions that we had to fill with long-term substitutes. It is important to us that the people we put in front of our students are given the tools to teach our kids. With that said, this year we included our long-term substitutes in our new teacher orientation. They go through the training our new teachers go through, and they, too, are given a mentor to receive support throughout the school year. They are observed and given feedback by our principals in the same manner other teachers are. In addition, we have started to grow our own within our division. One way we have started this is by partnering with a local university that offers a career switcher program. Through this program, Suffolk will pay for our paraprofessionals with four-year degrees as well as our long term substitutes who are interested in becoming a teacher to attend this program. In turn, we ask that those that go through the program work with our school division for at least three years.

Quite often now, candidates are not coming straight from a teaching program; rather, some are career switchers, some have degrees in the STEM field but had not originally considered teaching. According to the Joint Legislative Audit and

Review Commission (JLARC 2022) school divisions are hiring more teachers with provisional licenses. In Virginia, a provisional license is issued for three years. During those three years, the employee should take the courses listed in the letter they receive from the Virginia Department of Education. Once they complete all the requirements, they can then apply for a full license. During the 2021-2022 school year, provisionally licensed teachers increased twenty-four percent statewide. The goal in staffing teaching positions is to staff with teachers who are endorsed in their content area. However, according to the JLARC report, teachers not endorsed in their area more than doubled during the 2021-2022 school year. Unfortunately, the JLARC report also noted that school divisions lacked confidence in fulfilling teaching positions for the 2022-2023 school year. Over 52% of school divisions felt it was unlikely they would be able to fill all teaching positions with qualified teachers. With this, on top of teachers saying they are more likely to leave teaching, HR departments have a bleak outlook on filling vacant positions (Joint Legislative Audit and Review Commission, 2022.)

In spite of this concern, as director of the Human Resources Department, it is important that my team and I think outside the box. We can no longer just rely on job fairs hosted by colleges and universities. On July 17, 2019, former Virginia Governor Ralph Northam signed executive order thirty-six, which addressed STEM education. In this order, he noted, "In order to close achievement gaps, meet the growing economic demands for STEM-prepared citizens, and maintain the recognition as one of the best states for STEM education and employment, the Commonwealth must be strategic in how it prepares and educates young people and aligns to the rapidly evolving needs of employers. To accomplish this, Virginia must begin by creating a unified vision and adopting a set of shared goals to strategically inform how we prepare students for STEM jobs of the future" (VA Secretary of Education, n.d.). Science, technology, engineering, and mathematics (STEM) has been defined as: "An interdisciplinary (or transdisciplinary) approach that integrates knowledge from diverse academic disciplines into authentic problem- or project-based learning experiences as related to instruction in STEM content areas. Each of these is embedded within the scientific method and engineering design processes, as well as 21st-century skills" (Basham et al., 2010, p. 11).

Understanding what STEM is helps in the hiring process. When selecting candidates, we need to look for teachers that teach STEM through a hands-on, critical-thinking approach that considers how the student learns. Teachers must also realize that all students benefit from STEM education. STEM allows students to think at a higher level, be creative, and problem solve. These skills will benefit them no matter what their future interests are (*What Is STEM Education? | Britannica*, 2023.)

Due to teacher shortages, we have changed our approach for attracting, recruiting, and retaining teachers. First, we have developed a plan of support for those we hire. Through staff interviews, I have found teachers want professional learning that supports them in a way that allows them to take their professional learning to the students they teach. This year, in particular, I have worked closely with our professional learning coordinator to help develop a mentoring program that we will implement during the 2023-2024 school year. This mentoring program is a three-year plan that will provide professional learning experiences for our teaching staff that will build and enhance their skill set. Another important aspect of hiring is the retention of staff. At this point in the field of education, we can not out-recruit those who are leaving. For every new teacher we hire, there is a cost to the division. Although there is not one answer to retaining staff, we need to ensure our staff feels appreciated, respected, and heard. In my opinion, retaining staff is a key component to the success of any school division. The staff we have possess skill sets that have allowed us to depend on them to grow our teachers that are new to the field.

How does my role as the human resources director impact this? It is important to understand that to support the initiative to bring STEM education to students (our most important asset for our future), I need to specifically look for teachers licensed to teach in the areas of STEM. In Virginia, going into the 2022-2023 school year there were approximately 1,200 teacher vacancies. In Virginia according to the 2021 staffing and vacancy report, STEM teaching positions there were a shortage of about 18.5% (Virginia Department of Education, 2022.) Currently, there are 119 teaching positions that are currently vacant. Of those 119 positions ten are considered STEM positions. What have we done to address these concerns? During the 2022-2023 school year, we have addressed these vacancies in several ways.

To address this concern, we have done a few things. One, we have partnered with Universities which has allowed staff to get discounted rates on college courses they may need to meet the requirements of a provisional license or to meet the requirements for renewal of their licenses. For the last two years, we have offered a $2000 sign-on bonus and a $2000 retention bonus for critical shortage positions, including math and science. To receive the sign-on bonus, candidates must agree to work with us for the whole school year. To receive the retention bonus, they have to agree to work with us the following school year. Our department also applies for grants from Virginia that aim to recruit and retain teachers. An example of this would be the Recruitment Incentive for Public Education (RIPE) grant. Last year we had seven teachers awarded $5000 through the RIPE grant. This grant was for candidates hired during a specific time period that taught in teaching positions such as math and science. Last year we began to implement a referral program for our staff. If they refer a teacher, we pay our employee a $500 bonus. As long as there is funding to support recruiting, retaining, and referral bonuses, we plan to continue this.

I understand teachers will leave for various reasons. In saying that, it is vital we find out the "why" teachers choose to leave. We send out exit surveys to staff that leave the division to find out why they have chosen to leave. Sometimes the results are things we can not control, such as a family move or medical reasons. However, at other times we do play a role in the decision a teacher makes when deciding to leave or stay. For example, if a teacher does not feel supported, their administrator may have been able to give them the support needed had they been aware. This is why I have one-on-one conversations with staff who choose to leave, especially those who leave during the school year. I have found during these conversations, more often than not, they had not informed their administrator of how they were feeling. There have been times I know we can rectify a situation if given the opportunity. By taking time to hear teacher concerns and asking them for an opportunity to try to meet their needs, I have found we are able to retain them.

Although we have adjusted to teacher shortages, what we have done is not enough. We still have to find creative ways to attract STEM teachers. The state of Virginia continues to look at alternative pathways to licensure; however, in my opinion, there are still walls that make it difficult for people who are interested in becoming teachers to get a teaching license. Our Human Resources Department must be proactive, not reactive. We want new teachers, as well as our veteran teachers, to know we want them in our division, and we want them to want to work in our division. Potential STEM teachers often use social media. We need to communicate through social media, such as Twitter, LinkedIn, Facebook, and TikTok. It is important we stay abreast on what STEM teachers are looking for in a division. We need to collaborate with other divisions, not necessarily compete with other divisions. My department will continue to work with our finance department to ensure we provide feedback about teacher pay scales and share where we see support is needed (for example, providing staff to ensure teachers have a duty-free lunch.) I need to always make sure that I communicate staffing concerns with the superintendent and share with him solutions to ensure our students are getting the best from our staff. Finally, the staff in human resources understand our students need teachers that can help them be creative, think critically, work collaboratively, and have the ability to solve problems. It is our responsibility to find a way.

∾

References:

Joint Legislative Audit and Review Commission. (2022, November 7). *Pandemic Impact on Public K–12 Education.* Retrieved February 11, 2023, from http://jlarc.virginia.gov/pdfs/reports/Rpt568.pdf

VA Secretary of Education. (n.d.). *Stem Education Commission.* VA Secretary of Education. Retrieved February 9, 2023, from https://www.education.virginia.gov/initiatives/stem/

Virginia Department of Education. (2022). *Staffing and Vacancy Report*. Virginia Department of Education. Retrieved February 14, 2023, from https://p1pe.doe.virginia.gov/apex/f?p=352:1:15642680522386:SHOW_REPORT::::

What is STEM education? | *Britannica*. (2023). Encyclopedia Britannica. Retrieved February 4, 2023, from https://www.britannica.com/video/193418/overview-STEM-education

About the Author

JESSICA AVERY

Born In Arizona, Jessica Avery, was raised in a military family which allowed her to live in several different places throughout her public education journey. At the age of twenty-one, she joined the Navy where she served for five years. After leaving the Navy, Mrs. Avery finished college and decided to become an educator. Mrs. Avery began her career in education in Suffolk as a Special Education high school teacher in 2004. In 2006, she became a 4th grade teacher for five years. A servant leader at heart, Mrs. Avery felt she could have a positive impact on students and teachers by becoming an educational leader. In 2012 she was named Assistant Principal, and in 2015, she was named the Principal, where she remained for six years before being named Coordinator of Employee Relations. In March 2022, Mrs. Avery was named Director of Human Resources.

During her tenure as Principal, she built a strong leadership team and led her school from being a school in focus to a Fully Accredited school. As the leader of a school that is a model for school improvement, Mrs. Avery was asked to present at Virginia Association for Supervision and Curriculum Development. In addition, she presented to the Suffolk leadership team sustainable strategies for school improvement, using data to drive instruction, and effective use of professional learning communities. Building relationships is her passion and, more specifically, identifying students that are underserved and finding a way to become proactive to help them succeed in school. Through her leadership, she has worked with other educators to help them become educator leaders.

Mrs. Avery received her Bachelor of Arts in Criminal Justice from St. Leo University, her Masters in Elementary Education from Regent University, and her Ed.S in Administration and Leadership from Regent University. Currently, Mrs. Avery is working toward her Doctorate in Educational Leadership and Policy Studies at Virginia Tech, with her dissertation interest being Restorative Justice in Education.

Section V - The Future

THE FUTURE

How often in society are decisions made for students without obtaining the student perspective?

Our experiences in Suffolk Public Schools show that if students have a voice in decisions that affect their educational journeys, they will have more buy-in.

We wanted to ensure that the student voice is a part of this book, and we believe that every reader can learn a little something or two) from reading what our students had to say.

ONE

The Senior Perspective

Raven Cooper & Alexis Perkins

Raven Cooper

HELLO, MY NAME IS RAVEN COOPER, AND I AM A SENIOR AT Lakeland High School. I have known, just about my whole life, that I wanted to be a veterinarian and pursue a degree in biology. Being able to take both AP Biology and AP Chemistry has really helped me both get into the universities and majors that I want and also to get a glimpse into what college courses I like and ones I do not. I have learned that I enjoy biology, and taking AP Biology with Mrs. McGee was the class that solidified my decision to pursue a degree in biology. On top of being a wonderful teacher, she had lots of great tips and advice for me that helped me to make the decision to pursue biology this fall. I have also learned this year that I don't like chemistry as much as I like biology. My teacher, Ms. Jacobs is fantastic, but I just don't like the material. However, I am so grateful that I am taking this class because it is going to give me a huge head start when I do go to college and have to take chemistry. Knowing that SPS offers AP classes that give you a glimpse into the college world helps you narrow down what you like and don't. It also really prepares you for the difficulty and intensity of certain college classes.

I think adding more animal-science-focused classes would be super beneficial and exciting for a lot of the students. Animal science is an ever-growing field, and only having the vet-tech program at The College and Career Academy at Pruden (CCAP) isn't enough. Not everyone is able to go to CCAP and I believe that if there were more animal-science classes, it could really benefit some of the people that are looking to become a veterinarian and/or work in the animal science field. Along with SPS not having many animal-science classes, I believe that adding some other specialty classes such as botany, astrology, or forensics could be really fun. I know that it can be hard to believe that someone would want to take those types of classes, but having them as options could really help spark someone's interest and help them find their career. Another thing that could really help enhance the SPS STEM program would be the incorporation of internships. Internships look absolutely amazing on college and scholarship applications when it comes time for that. I think making it a program where you don't have to be in the Biomed program in order to qualify would be amazing. I believe that the internships would be best if done in the summer since it would be hard to have internships during the academic year. But allowing the students to have this opportunity to be interns would really help them to see if pursuing a certain career is what they want. It is one thing to think you want to pursue something, but it is a whole different ball game when you actually get experience in that certain field. I know for me, I always thought that I wanted to work with small, domestic animals when I was younger. However, as I grew up, I got to visit and talk to small animal veterinarians, and I realized that it was something that I really didn't want to do. Working at a horse ranch and being able to be so hands-on with such large animals really helped me to make the decision that I want to be a large animal veterinarian. Having that hands-on experience is something that I really think would make SPS stand out among the other school districts in Virginia. Having internships at hospitals, animal clinics, animal shelters, nursing homes, engineering jobs, etc, are just some general ideas that could benefit future school years. To me, SPS has done a great job in making sure the AP science teachers that we do have are all amazing and really do want the best for their students. I also like how they have made it so that those two AP science classes are double-blocked so the students really do have as much time as possible to learn the material and really apply it in the best way possible. One problem that I do see is that since these AP science teachers also have to worry about general ed science classes, they can be overworked and tired by the time the AP class rolls around. Although I do love and enjoy the double-blocked class, if the double block was in the middle of the day instead of the end, I think it would benefit both the students and the teachers. Having to do your hardest class at the end of the day after you have already had a full day of challenging classes can be very difficult and often, you aren't able to take in the information as well as you need to. I believe that allowing those double blocks to be in the middle of the day could help make those AP science classes more appealing to students and easier on the teachers as well.

I am honored to be able to call myself a Lakeland Cavalier, and I am very grateful for all of the opportunities that I have been offered while at this school. It is my hope that even when I am off and graduated, the STEM program will continue to grow and improve for many generations to come. GO CAVALIERS!

Alexis Perkins

Life as a biomedical science student is not always easy. However, my love for science has always been there. When I was in the sixth grade, I had already developed the mindset that I wanted to go into the medical field. With that mindset, I flourished in all of my science classes. When I was in the seventh grade, I was also put into the advanced science group so, I was ahead by a year. Being put into that advanced science group had its ups and downs, but overall it was an amazing experience for me and it was a great idea to even have offered that to me. However, this was only the beginning.

Moving on into my eighth-grade year, I learned about the Biomedical Sciences program offered at Lakeland High School. As soon as I heard about it, I immediately made it my mission to try and get into the program. I worked extremely hard to keep my grades up, and I even still remember how excited I was that a program like that was even available to me in the first place. Obviously, I decided to apply and later found out that I had gotten accepted into the program. Once accepted, my journey of furthering my education in science began.

My freshman year was kind of rough. It was my first year in high school, and trying to adapt to a new environment while trying to navigate through my first year in the program was challenging. However, my Biomed teacher is phenomenal, and she worked patiently with me the entire way. To this day my favorite moment from freshmen year was when we did our first lab and we extracted DNA from strawberries. From that moment on, I just knew that this program was the one for me. Unfortunately, halfway through my first year COVID-19 had hit, and I didn't get to finish. This was upsetting to me because I feel like there was so much more to learn that I, unfortunately, didn't get to experience.

My second year of the program was all online, so I did also miss out on a lot of the human body systems portion of this program. Even with having to miss a huge chunk of the program due to COVID, I still went into my third year, my junior year, with high hopes. I was eager and excited to get back into the lab and into the classroom to start learning again. Going from being out of the classroom for almost two years to back into the classroom was definitely challenging, but junior year in the program was definitely my favorite. We did so many fun labs that not only taught me new science knowledge, but also exposed me to lab techniques that will be very helpful to me when I get into college. One of my favorite labs from this year was the antibiotic lab. In this lab, we tested the resistance of e.coli against different types of antibiotics. I felt like it was a really cool experience, and I learned so much from the lab.

Finally, I entered my senior year. All three years leading up to my senior year were what prepared me for just how rigorous the fourth year is. The major part of senior year is completing a capstone project. It is very time-consuming and requires a lot of work and research. I feel like this capstone project is almost preparing me for college because I have to think more in-depth about stuff, and I'm having to put in twice the normal amount of work. That's one thing I really enjoy about this program. It's not just an easy course; instead it challenges me, and I feel like once I get into college, I'm going to be prepared.

Overall, the Biomedical Sciences program is truly amazing. The teacher, Sarah McDonald, has helped guide me in ways I didn't even think were possible. Without her, I feel like I wouldn't have had the experience I had. So I truly appreciate her and everything she has done for me. My love for science continued to grow throughout the entirety of the program, and it has helped shape me into the person I am today. I definitely feel like SPS should offer more STEM-related programs if at all possible because I feel like they can be very beneficial to those who plan on going into careers related to STEM. However, I am continuously thankful for even having the opportunity to experience all the technology and knowledge that comes with being in this program. I never imagined myself being where I'm at today and having this strong of a love for science, but this program really helped me find myself. At the end of the day, the Biomedical Sciences program will always have a special place in my heart, and it will continue to motivate me as I further my studies in the medical field and the science world.

About The Authors

RAVEN COOPER & ALEXIS PERKINS

Raven Cooper is a native of Suffolk, Virginia and is a Senior at Lakeland High School in Suffolk, Virginia. She intends to study Biology – Pre-Med at Texas Tech University when she enters college in the fall of 2023.

Alexis Perkins is a native of Portsmouth, Virginia and is a Senior at Lakeland High School in Suffolk, Virginia. She intends to study Pediatric Nursing at Radford University when she enters college in the fall of 2023.

TWO

The High School Junior Perspective

Alexis Griffith, James McFarland, and Ramirsjae
Young

Alexis Griffith

I'VE WANTED TO PURSUE A CAREER IN SCIENCE, SINCE I WAS IN my seventh-grade physical and life science class. When we learned about genetics, I was so amazed by all the different aspects of DNA. So, when we were handed pamphlets about Suffolk's specialty programs, I immediately picked one for the biomed program and didn't even look at the pamphlets for the other programs. I decided to apply because I thought it would be beneficial for the future, especially with colleges, and because I was hoping that I would learn a lot from the program. This program has taught me a lot so far, and I hope it will continue to provide some interesting knowledge. Another reason I applied was for all the different opportunities the program would offer. Visiting Operation Smile and learning about their mission was amazing and something that I would never have been able to do if I wasn't in the Biomed program.

I am very excited and nervous about my senior year. I hope to go on more field trips or internships during my senior year. From watching the seniors this year and last, I'm excited about all the opportunities that will be given next school year. It was great to watch the seniors' capstone projects and see how different each of their projects were. One group tested the connection between different scents and memories, which I thought was fascinating. It made me wonder what my capstone project would be. I also hope to be able to use the Anatomage Table, a computer that contains a library of cadavers that can be used to observe different body systems and has a dissection tool, more during my senior year. It would be nice to have more lab time during senior year. As a senior, I hope to have the opportunity to have an internship with a forensics unit of the police department or be able to shadow doctors at a nearby hospital.

James McFarland

There are multiple reasons why I applied to the biomedical science program at Lakeland High School. The first was my heavy interest in anything science related. I loved being exposed to knowledge, especially regarding the human body, specifically anatomy and physiology. After researching a little more about the program, I knew it was something that I wanted to participate in. I was also excited about how each year would focus on a different aspect of the science field. For example, the first year's curriculum had portions that consisted of forensic science, while the second year centered around the different structures of the body and its physiology. Another reason I decided to apply was that I knew how the completion of the program would look on my applications when I began looking at colleges. The rigor of the biomedical program and its connections with other schools, advertised by my program instructor, separates it from other higher-level classes. Each year becomes more challenging, and the skills you learn continue to build upon each other.

I'm very excited about my senior year and all the changes that come with it. From what I've observed, the senior class has greater independence in things like labs and/or research compared to the underclassmen, which is something I look forward to. Another change I'd like to see, and have already seen from the current seniors, is the opportunity to use our creative minds. For example, one presentation that I attended from the senior class involved creating toxicology reports and gross dissection results of a fictional patient to determine his or her cause of death. They are only given one word as a guideline, examples being "homicide" or "accidental", and are able to fabricate any scenario they would like. This made me very excited as I've always wanted the opportunity to branch out like this and see how creative I can really get. Other opportunities that I'm looking forward to are internships. An internship at a physical therapist's office would be ideal as it'd give me the experience I need to make future decisions regarding what occupation is best for me. Shadowing under a physical therapist is also something I plan on doing before I attend college to get an idea of what their "day in the life" looks like.

Ramirsjae Young

When I was younger, I always found science classes interesting because of the experiments and hands-on learning. The human body and dissection of animals have always been most intriguing to me as I loved to learn about how things function & why we are able to do the things we do, and by joining the Biomed program, I'd hoped to delve deeper into that subject. However, now that I'm in it I've realized that Biomed dived way deeper than I had originally anticipated. One interesting thing I learned within my time in the Biomed program is how the body maintains homeostasis. It does this through various feedback loops like sweating, which is what our body uses to cool us down; our muscles contract and shiver to heat us up.

One thing I look forward to seeing in the future of this program is going on more eye-opening field trips like the one to Operation Smile last year. When we were there, the staff told us about their various expeditions to help children get surgeries to fix their cleft lip or palate in order to give them a better life. That experience inspired me and made me appreciate the biomedical field even more. Another thing I look forward to is doing more unique labs for my capstone in my senior year. Last year I got to see their capstone project and the different aspects of it. For example, one group tested the efficiency of our school's cleaning products/methods at killing bacteria, and after seeing this and the other capstones, I couldn't wait for the time when I could do my own.

Another thing I look forward to in my senior year is the incorporation of internships into the Biomed program. This is because the internships allow us to apply what we have learned and even learn more about different biomedical fields. Internships I'd most like to see are working with nurses or doctors in hospitals and working with forensic scientists at labs. Even if we were only watching, it would be an eye-opening experience. It would allow us to connect what we've learned and how to put it into practice in the field. So for those reasons, it would be beneficial and educational to include internships in the Biomed program.

About the Authors

ALEXIS GRIFFITH, JAMES MCFARLAND, AND RAMIRSJAE YOUNG

Alexis Griffith is a native of Suffolk, Virginia and is a Junior at Lakeland High School in Suffolk, Virginia. She hopes to study Cellular Biology at Vanderbilt University when she enters college in the fall of 2024.

James McFarland is a native of Lynchburg, Virginia and is a Junior at Lakeland High School in Suffolk, Virginia. He hopes to study Biology and Kinesiology at the University of Virginia when he enters college in the fall of 2024.

Ramirsjae Young is a native of Patterson, New Jersey and is a Junior at Lakeland High School in Suffolk, Virginia. He hopes to study Engineering at Old Dominion University when he enters college in the fall of 2024.

Section VI - The School Community

If Suffolk Public Schools is serious about becoming the premier school division in the country, it also means that community partnerships and community support are paramount. It is necessary to foster sustainable partnerships with local businesses, community leaders, and parents. We need their perspective in order to learn what the locality needs and also to make sure that there is a combined effort to support the level of service required for our kids.

ONE

Serving the STEM Community

Dr. Rodney J. Brown

IN A WORLD THAT IS EVER-CHANGING, IT IS MORE IMPORTANT than ever before that our students are prepared to use knowledge and skills to solve problems, make sense of information, and know how to collect and evaluate information to make decisions. These abilities are what students acquire in a rich science, technology, engineering, and math (STEM) curriculum. If we want our students to become future leaders and solve some of the challenges of tomorrow, competence in STEM fields is essential. Schools and the communities where students live must ensure they have access to quality learning environments. A child's address should not determine nor limit students to a high-quality STEM program.

School divisions throughout the country are engaging students in various STEM opportunities to equip students with the necessary 21st-century skills. Too often our community prepares for the future by building industries to expand employment opportunities but does not prepare the workforce that is needed for these 21st-century jobs. It is the goal of all STEM programs to engage students and offer them real-world skill development. All students deserve to be exposed to STEM to reach their full potential, but too often in many divisions, there is a STEM opportunity gap among students. Schools and communities have to share the common goal to eliminate barriers for students that contribute to the gaps in science education. Many classroom teachers, families, and communities need more support to create and implement high-quality STEM learning experiences. As educators and community leaders, we have a responsibility to provide all students with opportunities and paths to our future workforce, and this is especially important for those who are underserved and underrepresented in STEM. Students need technical knowledge, as well as critical thinking, problem-solving, and analytical skills. These skills will prepare them for the jobs of the future. But it's not just the school division that should be thinking about how best to engage students.

In partnership with the community, it is vital that divisions build a network of Family STEM Communities. Family STEM Communities allow adults and children to participate in explorations, interactions, and conversations that are learner-centered. These communities bring together early childhood programs, schools, community centers, and libraries to improve the quality of teaching and learning in STEM. These partnerships will serve educators and community partners together to support students and their families who have historically not had access and equitable opportunities to engage with high-quality STEM experiences. Too often, this results in students' inability to see themselves as STEM learners.

So, what can communities do to support STEM in schools? When we look at the city of Suffolk and its growth over the past ten years, we see several new housing developments and an expansion of major companies choosing our city. Amazon, *Virginia's first Amazon Robotics Fulfillment Center*, is just one example of a company expanding to Suffolk. The center is filled with hundreds of robots to assist workers to do their job. The city also started its 2045 Comprehensive Plan. The title for this plan is *Suffolk 2045 - Connecting Our City, Shaping Our Future*. Suffolk 2045 provides an opportunity for everyone who cares about the City of Suffolk to share their ideas about the future. This comprehensive planning initiative will be used to identify long-term goals and desired policies, programs, and projects for a range of important topics. As the city begins to talk about shaping our future, we will have to discuss STEM experiences for our students. We cannot prepare for our future if we do not invest in the people responsible for making this vision a reality. Our students today are the ones that will need the skills for Suffolk to function and thrive as a city in 204. The year 2045 is a short 22 years from now. That means that every student in Suffolk Public Schools today will potentially have graduated from college or spent at least nine years in the workforce or military. These are the individuals who will be shaping our city's future. So as more companies like Amazon choose the city of Suffolk, it will be wise for this community to prepare the workforce that can support these 21st -century jobs. Therefore, cultivating a workforce that is ready to step into these STEM-focused jobs is vital. Suffolk 2045 gives everyone who cares about the City of Suffolk a chance to share their ideas about the future, which in turn gives the community a chance to engage in a conversation about STEM. In this planning, having conversations with our school division is also important. If the future of this city is currently in our schools, the city and school division has a responsibility to work together to

ensure that our students are prepared for future expectations.

In addition to supporting our schools, the city must also invest
in organizations like after-school programs, city community
centers, and libraries. After-school programs extend opportunities for students that may not have access to STEM
resources. Students spend about 80 percent of their waking
hours outside of school, while many of them are alone after
the school day ends. High-quality afterschool programs also
promote positive youth development and offer a safe space
where our students can explore their potential. Some examples of these after school programs are Parks and Rec, Boys
and Girl Club, and faith-based organizations. Effective after-
school programs also can provide a STEM-enriched learning
setting that brings a wide range of benefits to students, families, and the community. These programs also can support
social, emotional, cognitive, and academic development,
reduce risky behaviors, promote physical health, and provide
a safe and supportive environment for children and youth.

Community centers are also vital if this community wants to
stay on the cutting edge of what youth and teens love to
explore, such as video animation, 3D printing, and robotics.
Extending these opportunities to our city's youth can greatly
increase access to STEM education. Community centers are
also vital for a healthy community. These centers provide
opportunities for students and families to interact with other
residents. They hold the key to many benefits that enhance
the quality of life in this community. Regardless of race,
ethnicity, gender, age, or socioeconomic status, there are valuable programs at our community centers for little to no cost
to our city. A few examples of these valuable programs are
fitness classes, afterschool programs, or job training. Ultimately there should be something at our community centers

for everyone. However, to assist with preparing our community for the 21st Century, this city must invest in the resources offered in these centers. These centers also provide resources and opportunities for students that are not yet of school age. These STEM opportunities support cognitive development, build their brain capacity, and develop skills such as problem-solving before they start school. These are the skills every child will need to successfully work in the 21st-century. Lastly, libraries are progressing to become places for informal STEM education, providing programs and exhibits that inspire children and adults, and support them in their pursuit of lifelong STEM learning. Historically, libraries have served diverse populations of all ages and backgrounds and can be found in nearly every community across the United States. A library's STEM programs should be designed with the goal to inspire lifelong learning through inquiry and play. Traditionally, libraries have been a place of books and a place to develop reading literacy skills and establish a love for literature. When it comes to STEM, libraries have focused on reading about STEM topics rather than doing and practicing STEM. They offer books that include information in a variety of subjects (science, mathematics, psychology, computer science, technology, engineering, social science). All of these subjects can be enhanced through learning by doing. Informal STEM learning environments such as libraries can provide access to this type of thinking in the form of interactive learning activities. Members of the community value the role of public libraries in their communities, both for providing access to materials and resources and for promoting literacy and improving the overall quality of life. Community leaders must consider location and accessibility when building libraries.

Schools have traditionally been the place for our children to develop skills to solve problems and think critically. But schools are not the only place in this community where education takes place; they should also exist within our community's learning environment. Schools, businesses, libraries, and the home all make up a community's learning environment, and these places are designed to support the development of STEM education. It is no secret that employers are looking for workers with STEM skills and digital literacy. Nearly every career choice for our students will require technical and problem-solving skills. The foundation of any flourishing community lies in providing accessible advancement opportunities for its citizens. It is the responsibility of the community to ensure all children have access to the tools they need to secure sustainable, living-wage jobs. Essential to achieving this is early exposure to high-quality, hands-on STEM learning experiences that engage and inspire students. Together, the community and schools can ensure we are enhancing the potential of our students when it comes to STEM. That starts with investing in our community. The community must get to know the students in our schools and neighborhoods. Find out what they need for success, and connect them with resources and youth programs to help them get there. We need students of all backgrounds to contribute and participate in meeting the needs of a 21st-century workforce. With equitable access to opportunity and engagement, our students can build a foundation for a promising future.

About the Author

DR. RODNEY J. BROWN

Rodney J. Brown, Ed.D., is a former Special Education teacher, teacher mentor, professional developer, athletic coach, and building-level leader. He has over 20 years of experience at the school division level. Rodney currently serves as the Chief of Administrative Services for Suffolk Public Schools and is the Vice President of the Virginia Association of School Personnel Administrators (VASPA). He has presented at school, district, and state levels.

Rodney is a husband and father of two beautiful girls. He received a full-ride athletic scholarship for football from Norfolk State University, where he obtained a Bachelor's Degree in Psychology. Creating opportunities for students and staff is his passion. Rodney also has a master's degree with an administrative endorsement from Norfolk State University, and a doctorate from The College of William & Mary in educational policy, planning, and leadership.

TWO

STEM: The Impact of STEM in the Community and Community Support

Major David Miles

In 1951, Langston Hughes, in his poem "Harlem," asked the question of what happens to a dream deferred. Does it dry up? Fester? Does it become rotten, crust over, sag from its weight, or explode?

As a teen reading that poem, which I still hold dear, I tried to imagine his words. As a student, I always preferred words over numbers because my mind could make words come to life. As an adult, I now understand that numbers in the minds of a scientist, an engineer, and a mathematician have just as much potential to come to life.

Growing up, my mother was a lead computer operator working in a civil service position for the Department of the Navy. The times that I could see where she worked were amazing to me. I knew this place was special because I had to pass through a special security station to enter the building and security doors to enter each room. As an elementary school-aged child, it was hard to understand everything that was happening there.

In her work area, the rooms were always cold, and each room was packed with giant computers. There were computers everywhere. There were computers against every wall. There were computers divided into aisles, and the sound of the hum from the machines was loud. As a child, I thought there had to be hundreds of computers in each room.

As you are reading this, I'm sure your mind is picturing a room full of desktop computers or maybe laptops. You might even think there are tablet devices spread around on table-tops. Nope, not so. These machines were IBM magnetic tape drive computers. They stood over six feet tall, two feet wide, and four and a half feet in depth. They were built to store information on reels of magnetic tape. These reels were 10.5 inches in diameter, about an inch wide, and held 2,400 feet of tape.

[That is a lot of tape. Imagine this: you have a friend hold the reel and stand with their feet on the white line of the end zone while facing the opposite end of a football field. You grab the beginning of the roll of magnetic tape and hold onto it as you run to the opposite end of the football field. There is enough magnetic tape on one reel for you to run from end zone to end zone eight times!]

My mother and her coworkers would type commands on the keyboards that were connected to small monitors. The monitors at this time were not colorful like they are today. The screens were usually black, and the letters were either white or green. In addition to the letters, they also included slashes, colons, and other characters to give commands to the computers. There were shelves upon shelves of the reels stored on specially designed cabinets in the room. It was also a part of their job to exchange the reels on the machines throughout the day.

Just imagine a room full of these machines the size of refrigerators. Not to mention the room felt like a refrigerator. When I close my eyes, I can still feel the chill of the air and hear the hum of the machines as they worked.

Today, technology has advanced farther than anyone back then could've imagined. A one terabyte (TB) thumb drive can hold more data than one room of those computers. A smartphone can do more calculations and hold more data than a room of those machines could do in a day. Someone back then had a vision, or a dream, of improving beyond what was currently possible. Yesterday's ceilings have become today's foundation.

Years later, my mother tried to convince me to change my major in college to something that involved computer science. At the time, the government was hiring college graduates with degrees in the computer field and paying them better than they were paying my mother. Needless to say, I didn't listen. I graduated with honors from Norfolk State University with a Bachelor's of Science in Mass Communications and later began a career in law enforcement. Though my life has taken a path I never imagined it would take, and in a field outside of my degree, I credit my education as the key that unlocked the opportunities for my future.

In July 2003, during a speech commemorating the launch of a program to improve education and health care in South Africa, Nelson Mandela said, "Education is the most powerful weapon which you can use to change the world." Technology has caused the world to change in ways and at a speed never before seen in human history; however, education, and its influence, has been the catalyst for technology's growth.

As a child, then a teen, and a young adult, I changed my mind more times than I could count about what I wanted to do as a career. I wrote poems but couldn't rap or sing. I wasn't star-athlete material. Math wasn't my strongest subject. Still, in hindsight, the natural gifts I did have offered me more opportunities and options than I was aware of. The problem I had, and I'm sure others today may face, is not having a broad enough view of the landscape of options the world at the time and the future offered.

My father was born in 1947, and was the eighth child of the twelve. For my father, his landscape was a lot different when he grew up. He was born in the rural eastern shore of Maryland, in a small, two-bedroom house surrounded by farmland. He and his eleven siblings grew up sleeping head to foot in two beds, split between the boys and the girls. The children shared clothes, which included using patches to cover holes and using pieces of cardboard to cover a hole in the bottom of their shoes. To feed their family, my grandmother would gather wild greens that grew by the roadside, and my grandfather would slaughter the livestock he had in the backyard. In addition to going to school, he and his siblings would work in the neighboring farms and production plants doing odd jobs, including harvesting crops, cleaning chicken coops, as well as labeling and filling canned vegetables.

He once told me that he only saw two ways of starting a career to move outside of rural Maryland, either become a teacher or a minister. During that era, there were not many schools to choose from for minority students. When my father finished high school and looked to further his education, there were no student loans available to help with tuition. In order to pay for school, he worked, and his older siblings that had finished college and begun careers sent money home to help the next set of students get their education. Of the twelve children, one died from illness at a young age, the other eleven all received college educations.

As a child, his dream was to one day become a physical education teacher one day. To make that dream come true, he enrolled at Norfolk State Division of Virginia State College, now Norfolk State University. In 1969, he earned his degree and began a forty-year career as an elementary school physical education teacher for Virginia Beach Public Schools. After many years of being retired, he still hears from former students that thank him for his impact on their lives, some of which he taught the parent and their children.

Though my career path has not taken me directly down a path in STEM or as an educator as my parents did, I still draw from their experiences in hopes of helping others. Like a relay race, my parents and their generation ran their laps in the race and passed the baton to me and those in my generation. And now that it is my turn to run, I am holding the baton firm, not just so I don't drop it, but to make sure it is ready to be passed to the builders of tomorrow.

Like my mother, I want to help the builders of tomorrow have their hands in the technology of today and ask the question, "What can we do better?" I want to push them to take advantage of the wealth of educational resources that are available, so they can gain the knowledge and skills to answer that question.

Also, like my father, I want to encourage the builders of tomorrow to dream big. Get a vision of their dream. Draw it out. Write it down. Pursue it. What was once thought to be impossible was disproven when it became possible. What was once thought to only exist in our imaginations has become tangible. Right now, someone's dream is waiting to be released from their thoughts to benefit the whole world.

In the same way that educational opportunities have expanded, so too has financial support to assist students to reach their goals and achieve their dreams. For example, I am the president of the Suffolk Education Foundation, a non-profit organization that raises funds to help graduating seniors pay for college and awards grants to teachers for innovations in the classroom. We fund these projects not just through the generosity of our donors and supporters, but also due to their desire to provide opportunities for the builders of our tomorrows.

The Suffolk Education Foundation seeks to support opportunities for students to learn in new and innovative ways. This last year, we funded 16 instructional grant applications from teachers seeking funding for classroom innovations. There were many types of projects but the ones that excited me the most were STEM-related. Through these innovations, the teachers are allowing the students to be introduced to and put their hands on technology earlier, including opportunities like robotics, engineering and problem-solving. These early introductions are the planted seeds that can blossom into interests in the STEM field.

STEM disciplines walk hand-in-hand with the law enforcement profession. One of the most visible ways is the Crime Scene Investigator (CSI). Though they have received their fame on television shows, the job is a lot less glamours and a lot more important. The CSI's job is to go to crime scenes and collect all the evidence, no matter how small or seemingly unimportant, to help lead the investigators to the true culprit. CSIs use many scientific means of gathering, testing, and proving the collected evidence, and at times, in ingenious ways.

For example, the first CSI to work for the City of Suffolk Police Department was a Black woman, Joan K. Jones. She was hired by the police department in 1996 after working as a fingerprint examiner for the FBI. She processed all types of crime scenes, from burglaries and robberies to murders and more. Through the years of being on the front line, Joan has so many stories to tell.

One case that sticks out from her experience was an armed robbery at a fast-food restaurant. Three men entered the restaurant just before closing and shared two burgers, fries, and a drink. Five minutes before closing, the men threw away the remainder of their food and robbed the restaurant at gunpoint. A witness saw the men running away and called the police. Moments later, the police arrived, and even with the descriptions of the men, they were unable to locate them. When Joan arrived and gathered the information about what happened, she learned that the trashcans were emptied right before the men threw away their trash. That means the only items left in the trashcan belonged to the robbers. Now, normally evidence is placed in a paper bag to preserve it; however, this posed a bit of a dilemma because the food would rot and permeate through the bag, undeniably destroying any trace amount of DNA evidence. Instead, Joan used a plastic bag to protect the evidence. She quickly took the evidence to her office and placed it in a freezer, and the next day she took it to the lab to be tested.

As a result of Joan's meticulous actions and ingenuity, the lab was able to find fingerprints on the paper wrappers, the fry box, and the lid of the drink cup. Surprisingly, the lab was even able to get DNA from the saliva of the partially eaten burger. You could say the culprits didn't get to "have it their way" that evening. The fingerprints were compared to the files the police had in their database, but unfortunately no matches were found. However, not all hope was lost. When the DNA from the burger was entered into the system, the police were able to identify and apprehend the three suspects.

The success of the criminal justice system relies heavily on the groundwork done by CSIs, detectives, crime analysts, fingerprint technicians, and patrolmen. STEM careers are drastically improving and shaping the criminal justice system on a daily basis, creating alternate avenues for investigative techniques, evidence collection, and preservation, prosecution, and convictions. To be able to positively identify a criminal through the work of these STEM fields is critical in the criminal justice system as a whole. This positive evolution of investigative techniques allows the system to try the correct individual, eliminating the rate of false convictions and ensuring a just and fair trial. The justice system relies on the amazing skills and abilities of these STEM- based individuals to help ensure that justice is had by all.

Unfortunately, for many justice is just a dream. But, it doesn't have to be.

Today, there are dreamers who wish to be lawyers, business owners, servicemen and women, and more. There are also those who wish to shape tomorrow through medicine, science, public safety, and technology. Yesterday's dreamers made their dreams a reality for today. So, today's dreamers, you, can build us a better tomorrow.

In the poem "Dreams," Langston Hughes says, "Hold fast to dreams, for if dreams die, life is a broken-winged bird that cannot fly." Today, if the wings of our dreams are going to reach the future, education is the wind on which those wings will fly. The future is waiting. Opportunities are waiting. The builders of our tomorrow are traveling the paths of science, technology, engineering, and math. Today's leaders are waiting to coordinate the meeting between the builder's abilities and the future's opportunities.

About the Author

MAJOR DAVID MILES

Major David Miles is an honor graduate of Norfolk State University where he received a B.S. in Mass Communications in 2001. He began his law enforcement career in 2006 with the Virginia Beach Sheriff's Office, where he gained valuable experience in correctional and court operations.

In 2015, in order to serve in the community in which he resides, he transitioned to the Suffolk Sheriff's Office where he has faithfully served the city and citizens of Suffolk. Major Miles was appointed to Chief Deputy in 2019 and is the recipient of the **FBI-LEEDA** Trilogy Award. In order to build bridges between law enforcement and the community, Major Miles is a part of various community organizations and is the president of the Suffolk Education Foundation.

THREE

STEM: The Parent Perspective

Mrs. Tonya Byrd

As I write these words, I find myself in a state of amazement. My career path is not what I would have expected for myself, but I now realize that it is exactly where I was meant to be. It is the reason I was probably asked to contribute to this book. It is definitely why my husband and I have two children, our teenage twins, in Suffolk Public Schools' Project Lead the Way – Engineering Program at Nansemond River High School.

To explain why excelling in the STEM fields is so important to our family, you have to understand where that love of math and the sciences came from, and for me, it is not a traditional story. It began with a jar of coins and a pool table.

My uncle and godfather, Uncle Jeff, who I stuck to like glue, was a math teacher at Tabb High School in Yorktown, VA. In his spare time, he loved to play cards and shoot pool. I remember being four or five years old and carrying my jar of nickels to his house to play Deuces, Tunk, and 8-Ball. No jar, no games. For those not familiar with these card games,

Deuces and Tunk are about getting rid of the cards dealt to you by making pairs or a run of cards of the same suit. The person who gets rid of their cards first wins. This is how I learned about money, risk, probability, and statistics. Knowing that there are only four of the same number in a deck or that that you and your opponent may be waiting for the same card is key to winning.

I also learned about force, angles, and how every action caused an equal and opposite reaction. It may seem strange to think that his approach involved taking money from a child, but it worked. I had "skin in the game," so it was important for me to assess, practice, strategize, and implement. This was science at its core, and it was fun, cool, and exciting. For example, how and where you hit a cue ball determines its response. A harder straight shot makes the ball go straight while also stopping the cue ball from continuing along the path. Hitting it lower on its surface closer to the table causes it to spin, which, if done correctly, will have it roll back towards you once it connects with its intended target. And in the end, with lots of practice, I often walked away with more money than I started with.

That love catapulted from family games between a niece and her uncle to classroom application. There were three women from my K-12 years who solidified a career in STEM for me. Two of them were Mrs. Gibson, my 8th-grade science teacher at Lindsey Middle School, and Mrs. Mouton, my 9th-grade Earth Science teacher at Hampton High School. Ironically, Mrs. Mouton was also one of my mother's favorite teachers. The third was a friend of my aunt and fellow member of Alpha Kappa Alpha Sorority, Inc. I knew her as Mrs. Katherine, but the world knows her as Hidden Figure, Katherine Goble Johnson. It would take years for me to truly understand the greatness of these women. They were dynamic, intelligent, no-nonsense, caring, and commanded respect through poised grace. But, most importantly for a young African American girl, they looked like me. I am forever grateful for the imprints they left on my life.

And lastly, the summer before my senior year in high school, I attended Howard University's Upward Bound Math & Science Summer Program in Washington, D.C. While there, I worked with Dr. Emmanuel K. Glakpe, professor in the Department of Nuclear and Mechanical Engineering.

These experiences lead me to pursue a degree in mechanical engineering at Howard University, become a member of Alpha Kappa Alpha Sorority, Inc., and spend seven years interning for Virginia Power, now known as Dominion Energy, throughout my undergraduate and graduate studies.

This is my 30th year with Dominion Energy. I have said this a few times this year, and it still shocks me each time. If you are from Virginia, you know Dominion Energy. If you are not from Virginia or one of the other fourteen states where we have customers or operations, we are a Fortune 500 company headquartered in Richmond, VA. We employ over 17,000 professionals with a focus on providing reliable, affordable, and clean energy to nearly 7 million customers. We are leading the clean energy transition, and it is imperative that we invest in the communities in which we live and serve, and Suffolk is one of those communities.

My experience in the energy industry ranges from nuclear power to offshore wind. And while I agree that my parents, public school teachers, and educational curriculum were instrumental in my success, we cannot deny that the numerous external influences that were present were equally impactful and necessary. Students need all of these engagements to guide them to a successful career path where they are excited and energized, curious and thirsty for knowledge, and most importantly, wake up each day knowing that they matter and are contributing to a better tomorrow.

Tonya, and colleague Kaci Easley, accompanying a Dominion Energy offshore wind team performing routine maintenance at the Coastal Virginia Offshore Wind (CVOW) Pilot Project located 27-miles off the coast of Virginia Beach.

I would be remiss if I did not acknowledge that I am only one half of "Team STEM" in my family's household. My husband, who you will also hear from in this book, is a mathematician and educator. Even more than me, he is completely invested in education and preparing youth to be successful and contributing citizens to our world. We have very different approaches in how we provide experiences and learning opportunities for our twins, but it is absolutely the best way for them to learn what society and their futures hold.

Other than the fact that our twins, Trae Denise and Stenette IV, are both clever and funny, loving and kind, and competitive student-athletes, they are like oil and water. One is a planner, who prefers structure and is currently thinking about a career in medicine. The other, wildly free and most days just going with the flow, is still pondering life's path. Where one twin once included a stethoscope, blood pressure cuff, and first aid kit on her Christmas wish list, the other requested computer parts for his gaming system. Who knew water-cooled radiators for desktop computers were a thing? As you can see for yourself, even in their difference, STEM engagement still permeates.

We must give students creative and diverse opportunities to learn in all environments – in and out of the classroom. A successful path forward means exposing them to things they do not see every day. While I recognize my children's path was not exactly the same as my childhood experience, I used it as a model and stepping stone. The future of our youth depends on all of us – corporate, community, and schools – working together.

About the Author

MRS. TONYA BYRD

Tonya Denise Byrd is a native of Hampton, VA. She earned a B.S., Mechanical Engineering from Howard University and a M.B.A. from The George Washington University, both in Washington D.C. Her 30-year career with Dominion Energy has spanned several sectors, including nuclear engineering and training, customer service, and electrical distribution design. She now serves as Director, Community Engagement & Local Affairs.

Tonya serves on the board of The Cheryl J. Tyler Foundation, Fort Monroe Foundation, Hampton Road Chamber, Tidewater Community College Educational Foundation, and UNCF Virginia Leadership Council. She is a proud member of Alpha Kappa Alpha Sorority, Inc., Jack and Jill of America, Inc., and the American Association of Blacks in Energy.

Tonya lives in Suffolk, VA with her husband and teenage twins.

in

FOUR

STEM: The Business Perspective and Hireability

Dr. Jessica Johnson

"Those who are illiterate within the 21st century will not be those who cannot read and write, but those who cannot learn, unlearn, and relearn." -Toffler (1991)

CHATGPT. THERE. I SAID IT! JUST LIKE RALPHIE IN THE Christmas Story-put the soap in my mouth! At present, nothing that I can recall in recent years has been more educationally dichotomizing or polarizing than the inception of an artificial intelligence platform, such as ChatGPT, whose ubiquitous access to global information has presented conditions for such rapid societal and educational change. ChatGPT, or Chat Generative Pre-trained Transformer, is a natural language processing tool that allows users of any age to do a number of tasks such as text generation, generating story ideas, acting as a tutor for homework questions, and more. As an artificial intelligence agent, ChatGPT operates based on algorithm training from datasets to form what is called a deep learning neural network-programming modeled after the human brain. The simplest way to visualize ChatGPT as an automation tool is to compare it to the "complete the sentence" games we played as kids. But with powerful computing behind it. With ChatGPT becoming an almost instant global success, this creates challenges seemingly overnight for K-12 ecosystems. At the forefront of any educator's mind is the horrifying realization of, yet again, another technological shift to adapt to. Just like the "T" in the acronym STEM, the "T" in ChatGPT can put the terror on the face of any teacher thinking about all the countless hours and summers of curriculum and assessment development that are at stake. In reality, there are other, more powerful, AI tools besides ChatGPT. What frightens most is the game changing and rapid shifts that can reshape everything for an educator. Or. It can invigorate an educator to try a new type of technology.

For me, it was in the early 2000s. I was an elementary teacher attending a social studies conference wandering aimlessly in the exhibit hall when I saw it. It was beautiful, new, and in hindsight, huge and bulky. My first glimpse and demo of the capabilities of a Smartboard, and I was hooked. My classroom had old chalkboards lining the walls;I wanted and needed this. The visions in my head of the fun (and I guess learning) my students and I would have with a Smartboard brought me utter elation! The day came later that year when classrooms in my school actually received a few. These mammoth contraptions were never mounted to the walls but were on wheels so teachers could share across grade levels. I'll admit now that I probably dinged up multiple walls and door frames in attempts to wheel the heavy contraption down the hallways. The excitement was real in the beginning, and then it became cumbersome. I planned amazing lessons utilizing the Smartboard, only to have learned the morning I needed to wheel it down to my classroom it was either not working or someone else grabbed it before I could (even though I properly used the checkout system to reserve it). Technology is great. When there's ample access, systems, and training in place. Did my students suffer drastic learning setbacks because we were unable to use digital technology at times? No. In fact we often had more fun without devices than when faces were glued to screens. Technological shifts in education, as seen with ChatGPT, are viewed most often from a negative mindset. Yes, there are causes for concern over potential student cheating, negative implications when in the wrong hands, but what often makes us uncomfortable as educators also causes us to grow in our craft. What is often forgotten is that *technology*, as a concept, does not always equate to the utilization of electronic devices. Technology is the application or transmission of knowledge for practical purposes. It can refer to any tools, machines, objects, and other devices that are used to solve problems and improve efficiency. In the classroom, technology could be the glass beaker being used for a science experiment, the jars of multi-colored play dough being used to help students visualize Earth's layers, or simply just a paper and pencil. At one point, these tools were technological enhancements to learning.

Technological advances have always shifted the ways learners of all ages access information for knowledge acquisition. The dynamic demands placed on modern learners today exceed the need to regurgitate content knowledge, and instead emphasizes the importance of adaptation to new information and skill sets for both students and educators, known as the Learn-Unlearn-Relearn cycle. With each onset of new educational or technological advances encompasses an active "unlearning " outdated ideas, beliefs, practices, and behaviors facilitating to "relearn" or update one's knowledge and skill sets to stay relevant in an ever-changing world. This approach assists our students as learners in staying current to develop the critical thinking, problem-solving, and collaboration skill sets needed to be successful- aspects often considered to be the core foundations of STEM. Although a difference now, as opposed to yesteryears, may be the rate at which technological progressions occur. Each new mechanization introduced resumes the Learn-Unlearn-Relearn cycle. Change can be intimidating, though. I therefore giggle to myself as I reminisce on a meme from several years ago, see Figure 1, depicting how different generations view transformative changes. Insert gruffy voices and scoffs such as: "back in my day" or "these kids and their fancy technologies."

All jokes aside, I see this every day as an immersive learning researcher, developer, and learning engineer at Old Dominion University's Virginia Modeling, Analysis, and Simulation Center (VMASC). My passion and research are to assess, design, and develop immersive learning environments as integrative learning systems. STEM skill sets and emergent technologies are foundational elements in my everyday life. I may have physically left the classroom about five years ago; students and educators are always at the forefront of my work each day. I employ learning engineering domains and principles harnessing the capabilities of virtual reality (VR), augmented reality (AR), 3D simulations, and other advanced learning technologies to design/create engaging learning experiences across the lifecycle. Or utilize learning analytics and data science to assess and extract metrics from various learning systems, sometimes using biofeedback or neurofeedback devices. Often these projects require collaboration with professionals with varied experiences, and opinions, on utilizing digital technologies for learning or training. Their apprehensions or unwillingness to try a new approach inhibits project outcomes. New technologies, platforms, or mechanisms to streamline workflows are released seemingly each day. What makes emergent technologies, like ChatGPT, so domineering for educational communities are the multitude of ways they can threaten archaic methodologies of teaching and learning. Given the appropriate resources and efforts to research the utilizations, pros and cons of an emergent technology, these advancements can help transform knowledge acquisition for both students and educators.

While some may not keep pace, personally, with the advancements in emergent technologies, students often seize these ameliorations with reckless abandonment. These same students who latch onto the latest tech craze can adapt quicker to shifts in technology than, say, someone my age, but lack the comprehension of "where" to locate their file on a desktop computer. More importantly, they may have a unilateral understanding of tech devices, for example, virtual reality is only used for gaming and entertainment. I have witnessed similar instances with educators. Their fears or discomfort with new instructional practices utilizing emergent technologies hinder the potential learning experiences and outcomes for their students. There are a multitude of different ways these emergent technologies can be integrated not just within the classroom and aligned to state educational standards, but in combination as embedded systems. Often to believe it, one must see and engage with it.

Learning, UnLearning, & ReLearning with VMASC

A favorite aspect of my job at VMASC, besides my research and collaboration with colleagues, is STEM outreach to educators and students. Any workshop or professional development opportunity I lead encompasses engagement in real-world applications of emergent and integrative technologies. The goal is to make these experiences less "scary" for teachers and provide relevance to students.

A popular example is an engineering design challenge entitled, The Incredible Bulk (see Figure 2, Incredible Bulk EDC). In this challenge, teams are provided background information; they are marine engineers who are tasked with building special components of a bulkhead, the interior walls, on an aircraft carrier. Teams learn background information on what a bulkhead is and how it looks by visualizing the environment through virtual reality. Their task is to then replicate the components seen in the VR headsets using a 3D model. This model is interactive- they can manipulate it visually to see different angles or viewpoints. Most importantly, the 3D model helps them create and determine the measurements of where the physical components, 3D printed parts, should be placed on their bulkhead walls. Teams build 2 sets of piping systems, ladders, a hatch, electrical systems, and an HVAC system. To accomplish their task, teams must utilize STEM skillsets of collaboration and problem-solving to consider how to sequence the building and measuring of the parts. Teams are directed to signal they are ready for an inspection, like a real shipyard, where we use an augmented reality app to visually show teams if their component was built correctly and located correctly on the bulkhead.

The Incredible Bulk engineering design challenge provides students and educators an opportunity to understand how multiple technologies can be integrated for experiential learning outcomes. The activities are aligned to over twenty Virginia Department of Education Standards of Learning (SOLs) such as mathematics, computer science, language arts, and more! In addition, the engineering design challenge exposes students to digital transformations occurring in their local region and careers often underrepresented in STEM, such as skilled trades. More importantly, they are immersed in an experience to shift preconceived notions or apprehensions of how or which emergent technologies can be used in classrooms.

Shifts in STEM or the prefer-ments of new technologies require stages of learning, unlearning, and relearning for every educator. This still causes trepidation. Some worry that internet access in their classroom may hinder the experience for students, they lack training or  knowledge on how to use the devices, or they do not have the funds to purchase high-tech gear. I want to add to the phrase "It takes a village to raise a 21st-century graduate...*and teacher*". Industry partners, like myself, are here to support you! ALL OF YOU! We can help you tame the ChatGPT monster... and more.

～

References:

Ledes, A. (2014). Stop sharing this photo of antisocial newspaper readers. Retrieved from

https://medium.com/alt-ledes/stop-sharing-this-photo-of-antisocial-newspaper-readers-533200ffb40f.

Toffler, A. (1991). *Powershift: Knowledge, wealth, and violence at the end of the 21st century*. Bantam Books.

Tough, A. (1982). The other 80 percent of learning. In R. Gross (Ed.), *Invitation to lifelong learning* (pp. 153-157). Follett Publishing Company.

Van Eck, R. (2006). Digital game-based learning: It's not just the digital natives who are restless. *EDUCAUSE Review, 41*(2), 16-30. http://www.educause.edu/apps/er/erm06/erm0620.asp.

About the Author

DR. JESSICA JOHNSON

Dr. Jessica Johnson is a Research Assistant Professor and the Director for STEM and Educational Partnerships at the Virginia Modeling Analysis & Simulation Center for Old Dominion University. She is a Learning Engineer who uses applied learning sciences to assess and design immersive learning environments and content.

She received her PhD and Ed.S. degrees in Educational Psychology and M.Ed in Education/Curriculum Instruction Design from Regent University. She graduated from Edinboro University with a B.S in Cognitive Psychology. She has over 15 years of experience applying cognitive and educational psychology principles, learning engineering, and M&S approaches for utilization in education and training simulations. Her work and research encompass the applied assessment, development, and design of learner autonomy constructs in emergent technologies.

Dr. Johnson's funded grants and projects span collaborations with regional K-12 and higher education institutions, in addition to government, military, and private sectors. She also leads various STEM outreach and programming for audiences spanning K-higher educational partnerships. She previously taught as an educator in Suffolk Public Schools and Isle of Wight County Schools

FIVE

The Business Perspective and Hireability

Mrs. Talisha Anderson-Cheeks

MY LOVE FOR NURSING HAS BEEN ROOTED IN MY BLOOD FOR AS long as I can remember. "My little nurse" was what my family called me. I was five years old when I started organizing and administering medication to my grandmother. As a little girl I loved living with my grandma, although I had my own room at my parent's house. My grandmother, affectionately known as Granny, needed a lot of my assistance due to chronic health problems. Granny suffered from hypertension, diabetes, and coronary artery disease. I would administer her insulin shots, organize her medication in small pill pouches, and make her a cup of coffee every morning before leaving for school. At that time, I was only eight years old. My parents and friends were both in awe because I would rather be inside caring for Granny than playing outside. It was not just about caring for her; she spoiled me too. It was often only the two of us spending quality time together. We loved our lifetime movies, *The Golden Girls*, and *In the Heat of the Night*. As a child, I was very afraid of the dark, and being alone was my phobia. At my parent's house I had to sleep in my room alone, but at Granny's house, I slept in bed with her. I felt safe;I did not fear the dark being next to her.

My parents understood my love for being at Granny's house. On Fridays, my mom would pick me up to spend the weekend at home because I was also a daddy's girl, and my brother would spend the weekend at Granny's house in my place. With my fear of the dark, I would crawl into my parents' room once they'd go to sleep and stay until sunrise. I was often caught sleeping on the floor next to their bed with my doll and blanket. They would laugh about it, pick me up, and place me in my bed. My mom was a full-time housewife, and my dad was a long-distance truck driver; he would be gone three to four days a week. My mom was also a nurturer. She cared for my grandma while I was in school. She worked as a nurse aide part-time to elderly clients in their homes. I had a wonderful childhood; my family provided me with love, protection, and structure to be the woman I am today.

Tragedy happened in my life when I lost my mom at 13 years old. My mom underwent a surgical procedure called a carotid endarterectomy. The surgical procedure removes plaque from the two main blood vessels in the neck. This was supposed to be a simple procedure; however, it led to multiple strokes and, ultimately, an aneurysm. My mother was on a ventilator for one week at Sentara Norfolk General Hospital. After she was given a poor prognosis, my dad and grand-mother made the decision to remove the ventilator. The doctor's prognosis was a vegetative state due to multiple brain infarctions. At 38 years old, my mom lost her life. I remember standing at my mom's bedside as she took her final breath. The ride home from the hospital was dead silent. My dad told me years later what I said to him: "Daddy, I will find out what happened to my mom, and I will change the world." I had no idea what it meant at the time but, apparently, I spoke it into fruition.

I was a motherless child going through puberty. My mom passed a few days before my 8th-grade graduation at Forest Glenn Middle School. My high school years at Lakeland High School were amazing. My favorite classes were chemistry, biology, and algebra. I remember the day we dissected a frog. My classmates thought it was gross, while I felt fascinated by how the body works. I didn't know it then, but STEM always held a place in my heart. I had tons of friends and amazing teachers who showered me with love. Many of my friends from middle school knew my mom, and their moms would also shower me with love. The nurturing side of me was still there but I wanted to hang out with friends to distract myself from the reality of losing my mother. I met my first boyfriend at Lakeland. He had street credit, but I was still a bookworm. I lost my virginity to him and, eventually, became pregnant at the beginning of my senior year. My grandmother and father were extremely disappointed. I had plans of going into the Air Force to become a flight nurse, but that all changed when I became a mother. I let several people down, including my chemistry teacher, Ms. Ruffin, guidance counselors, and my principal as they encouraged me to be the best. I was very ashamed by my actions, so much so that I would wear large clothing to chemistry class to hide my baby bump. In my heart, I was another statistic.

I gave birth to my son July 29, 2002, one month after I graduated high school. I have never felt so lost. My father was heartbroken, and my grandmother was extremely hard on me. Do not get me wrong. They loved my son; however, my grandmother did not make it easy to consider a second child. I had to pay her childcare to watch my son while I worked at McDonald's. It got so tough that on some days, I had to give her my entire paycheck for bills and childcare. At 19 years old, I decided to get my own apartment with my son without any intention of ever returning home. I resolved to apply for nursing school, but assumed I had to become a nurse assistant first. I enrolled in a 12-week Certified Nursing Assistant (CNA) program at Paul D. Camp Community College while working full-time in the dietary department at Lake Prince Woods Retirement Home. My passion for nursing was always present but I needed to provide for my son as well. After completing my CNA program, I applied to become a unit secretary at Bon Secours Mary View Hospital. I left Lake Prince after five years of service to transition into my new role in the emergency department. I thought this was a creative way to see nurses and doctors in action. After working several months on the night shift, I decided to take my career to the next level and become a Registered Nurse (RN).

I applied for nursing school at Tidewater Community College and was accepted into their nursing program after several months of waiting. Nursing school was a challenge as a full-time single mom and full-time employee working the night shift. There were days I left work in the morning, picked my son up from the sitter, prepared him for school, and prepared myself for class at 9 AM. That was my pattern for two years until I graduated from nursing school in May of 2011. In March of 2010, I had lost my grandmother to lung cancer. I was at her bedside when she passed at home. I played a huge role in her medical care as I decided to move out of my apartment and back home with her after the cancer diagnosis. After graduating with my associate's degree as a Registered Nurse (RN), I took a role in the emergency department. My dream of nursing finally came true, and this was my foundation to my nursing career.

It was not hard to climb the nursing ladder. I had several hurdles to cross, but I never lost my passion to change the world. The start of the challenge was the loss of my mother. The next was the loss of my best friend during my first year of nursing school and, last, the loss of my grandmother during my senior year of nursing school. The three important women in my life who believed in me were gone, but they were never forgotten. I pushed myself every day to not give up and remembered they believed in me. I made an oath to live for them and change the world like I had promised. During my first year as a Registered Nurse, I was asked to be a Clinical Care Leader to my peers in the ER. In a leadership setting, the requirement is a bachelor's degree. Thus, I applied and was accepted into Norfolk State University Bachelor of Science in Nursing and graduated in December 2013 with honors.

I enjoyed leadership as an RN, but I felt like I could do more. I was a part of several committees in the hospital. The stroke committee was one that I enjoyed the most. It helped me understand how I lost my mom at such a young age due to a stroke. I was eager to see how this could have been prevented based on the risk factors she ha. One of the nurses I worked with graduated with a Master of Science in Nursing and became a family nurse practitioner. She began coming to work with a white coat and no longer worked as a Registered Nurse – she was a provider. Nurse practitioners are permitted to prescribe treatments, order tests, and diagnose patients. These duties are normally performed by physicians and are also implemented by nurse practitioners. I would sit down next to her and ask millions of questions about how to become a nurse practitioner. She willingly provided the information and I started doing my research on local schools I wanted to attend.

I applied for South University's nurse practitioner program and was accepted in June of 2016, exactly six months after graduating with my bachelor's degree. I worked full-time in the emergency department as a clinical lead while in graduate school for my master's degree. My second year of graduate school, I met my husband. We both worked extremely hard to build our lives as a blended family as we both had children from a previous relationship. After six months of dating, I became pregnant with our son. This time while being pregnant in school, I had the support of my husband and our family while pursuing my master's degree. I graduated with honors – summa cum laude from South University in December of 2016. I continued to work at the bedside as a Registered Nurse. I applied for a Nurse Practitioner role within Bon Secours in colon and rectal surgery. I interviewed for the new role and was offered the job as a Nurse Practitioner days later. I transferred to the surgical role in August 2016 as an NP. It was a new position for me and the physician I worked with. The transition was a challenge, but I had faced bigger hurdles. I did a lot of studying to master my fresh role as a provider. I spent a year in colon and rectal surgery before transitioning to a new role in bariatric surgery. I was very reluctant to take a role in bariatric surgery because it didn't feel like my calling. I really wanted a role in cardiology geared towards management of hypertension and diabetes. However, the role in cardiology did not work out, so I decided to take the bariatric position. This choice launched my career path to where I am now.

Bariatric surgery and supervised medical weight loss help people lose weight medically and surgically. It reduces the risk of life-threatening weight-related health problems that include heart disease, stroke, diabetes, and hypertension. I absolutely enjoyed that role at Bon Secours Surgical Weight

Loss Center. My passion for helping others and preventing illnesses combined with this role was everything I dreamed of. I chose a STEM-related field to make an important impact in the community. I had no idea that weight loss consulting was my calling. I worked alongside two amazing bariatric physicians who taught me everything about helping patients lose weight. I would often share with my husband how much I'd hear my clients at Bon Secours wish I could see them more frequently and how I wish I could provide those services. The patients were scheduled to see the provider at least once a month; however, for some patients that struggled a little more than others, they required more frequent visits. The patients that were evaluated on a more frequent basis were more disciplined and consistent with their weight loss than those with monthly visits.

During the pandemic, the bariatric surgical and non-surgical program was placed on hold. My husband decided to get the ball rolling with my business structure. One day I came home from work and the Virginia State Corporation Commission (SCC) documents with my business name, logo, and a startup check were on the table. He was tired of hearing me wish for things I was afraid to change. I had no excuse to keep complaining about working in Corporate America. I owe a lot to my husband, as he has been the mastermind of my business. Thank you, Kareem A. Cheeks for believing in my dreams! My last day as a Bon Secours Surgical Specialist was my grandmother's birthday, December 22, 2020, after fourteen years of service.

Goodbye, Bon Secours, and hello, First Choice Anderson FNP Health & Wellness Consulting LLC. First Choice Anderson was established in my home in January 2021, where I started with one client. My husband designed a room in my home as my office setting. While my business was in the beginning phase, I worked as a contractor for the city of Virginia Beach until it grew. After several months of taking on clients, I hired two part-time employees to assist with the patient load. We worked out of my home part-time via tele-health before moving to brick-and-mortar in March of 2021. I rented an office space from a good friend until I was financially secure to open my own practice. After six months of renting, I had enough clients to support my own office space. The space had two exam rooms, one office, and a reception area. Initially, I thought it was too big for the volume of clients I had. My goal was to stop at one hundred clients. Obviously, my vision was too small because we exceeded that number within a couple of months of opening my own space. We outgrew the space within four months and expanded to a larger space to support the volume.

About the Author

MRS. TALISHA ANDERSON-CHEEKS, FNP-BC

Talisha Anderson Cheeks is the owner of First Choice Anderson Health & Wellness Consulting in Suffolk, VA. First Choice Anderson provides weight loss services to patients in Virginia, Richmond, Maryland, North Carolina, South Carolina, Florida, and California to name just a few states. The business is growing daily. First Choice Anderson now has multiple providers and several clinical/non-clinical employees to support our clients. We see patients via tele-health and in-office on a weekly basis. We offer a maintenance program after the patient reaches their goal weight. First Choice Anderson's mission is to take a psychological/comprehensive approach to meet the needs of the clients on an individual basis.

The goal is to help each client reach a healthy goal weight that will resolve weight-related illness and establish a state of health and well-being. With my business and the programs it offers, I hope to honor the memory of my mother and grand-mother. I became and exceeded "the little nurse" title my grandmother always called me and sought to help many other people gain good and lasting health.

Comprehensive STEM™ Made Simple

The urgency of STEM skills and talent development

In the last 20 years, there has been a dramatic shift in the way we live and work, thanks to advances in technology, increased globalization, and the need for flexible work environments. There has been tremendous growth in STEM (Science, Technology, Engineering, and Math) jobs; however, over 800,000 jobs remain unfilled today due to the lack of workers with the skills needed to fill these positions. These numbers just represent the United States but the shortage of individuals who are [1]able to fill STEM jobs is a global concern. According to the Bureau of Labor Statistics, 13 of the top [2] 30 fastest growing jobs in the US are in STEM fields with growth of greater than 30% by 2030 . The reality is that *only 20% of US high school graduates are prepared for STEM careers or further* [3] *college education in STEM fields.* Moreover, socio-economic factors, race and gender remain strong [4] predictors of STEM career pursuits and are reflected in the persistent lack of progress in the [5] diversification of STEM careers. In 2022, the US

Department of Education released their priorities including the need to promote equity in student access, support a diverse educator workforce, meet social, emotional, and academic needs, and strengthen community engagement to advance [6] systemic change. Now is the time for school districts to invest in Comprehensive STEM™ strategies to prepare students for their future careers.

There are three problems that currently exist in our education system that contribute to the lack of STEM talent and skills needed to meet the demands of the workforce.

1. Lack of Access: All students do not have access to evidence-based STEM learning strategies, such as project-based or inquiry-based learning and design-thinking strategies that promote STEM skill development and career readiness.

2. Lack of Culture: Educators have not been trained to use evidence-based STEM learning strategies in their lesson plans, especially in an integrated or cross disciplinary way.

3. Lack of Community: Schools and Communities need to come together to facilitate partnerships that allow students to learn and develop skills in a real-world setting and allow employers to work with educators to ensure they are providing education that is relevant to changing workforce needs.

The changes necessary to promote the development of 21st Century skills and address the educational priorities and needs identified by the changing workforce can be accomplished through Comprehensive STEM™ education. Access, culture, and community are essential components of a comprehensive STEM integration initiative. Comprehensive STEM™ education provides access to integrated STEM experiences to all children early, often, and everywhere. Comprehensive STEM™ education requires a cultural shift in how educators teach students and curricula are developed. Comprehensive STEM™ involves building connections between the school and the community to promote opportunities for students to develop 21st Century skills through real-world experience.

Technological advances will continue to expand and the needs of the STEM workforce will change. As a result education must change and adapt to prepare students for the workforce. A 2019 working paper developed by the National Bureau of Economic Research demonstrated a rapid shift in STEM skills needed in the workforce from 2007 to 2017.[7]

> *"It is the new STEM skills that*
> *are scarce, not the workers themselves."*
> — *Deming, 2019.* [7]

Deming and Noray [7] show that the STEM skills that shifted from 2007-2017 reflect an increased need for Social, Cognitive, Character, Creativity, and Analytic skills among others. These skills have been the focus of project-based learning strategies (PBL), and inquiry-based or design-thinking strategies. Project-based learning strategies allow students to practice the skills of collaborative work, allow for creativity and critical-thinking, and problem-solving, while building resilience and character through trial and error. One of the first randomized controlled trials of students receiving a project-based inquiry science curriculum demonstrated that students achieved higher scores on science outcomes [8] compared to students who did not participate in project-based learning. Other studies suggest that PBL increases STEM skill development; however, there are still few studies that have systematically investigated the impact of PBL on a large scale, in diverse populations, across all grade levels. Still, the bulk of the evidence available suggests that PBL is a learner-centered strategy that can help to reduce the disparities in academic achievement, while building necessary skills for STEM workforce readiness, and helps to increase student motivation, knowledge, and self-efficacy for learning all of [9] which are associated with improved achievement outcomes. Unfortunately, there are few educators[10] who have the training needed to use PBL in their lesson plans.

Solution: Use PBL and Design-Thinking to provide STEM skill experiences to all students.

21stCentEd has developed a Comprehensive STEM™ solution that addresses challenges faced by school districts. Using a flexible, digital-learning platform that is accessible in-school, after school and out-of-school access to STEM experiences can be made available to all students. The platform provides learner-centered experiences that provide students the opportunity to practice 21st Century Skills through a design-thinking and project-based learning approach. These experiences can be accessed through a self-paced learning model or integrated into existing curricula. There are over 50 STEM courses currently available for students in grades K-12, with courses organized into learning suites that provide in-depth experiences in: Computer Science and Information Technology, Engineering and Robotic Technology, and Workforce Development, Entrepreneurship and Business. The student license gives access to not just one, but all the courses in the catalog. Since the license follows the child, community partners may act as an additional access point and would only need to provide their own mentors. A recent meta-analysis reports that the use of digital tools to enhance[11] STEM education produces achievement gains. The digital-technology platform provided by 21stCentEd provides a solution to the disparities in access to STEM experiences using a curriculum designed to provide over 1100 hours of STEM skills practice and experience.

ACCESS WHERE THEY FEEL MOST CONFIDENT AND SUPPORTED

50+ PROPRIETARY AND VARIED STEM COURSES/EXPERIENCES

1100+ HRS OF STEM AND CTE INSTRUCTION

SELF-PACED, BLENDED, OR INSTRUCTOR-LED DELIVERY

ONE STUDENT LICENSE PROVIDES ACCESS TO ALL COURSES

Additional Features

- Courses are project-based and provide practice in 22 different 21stCentEd Skills Categorized in STEM/CTE pathways
- Can be implemented using self-paced, blended, or instructor-led methods
- Students can build a digital portfolio including acquired badges and competencies Experiences range from 10 hours to full semester experiences
- Customized course creation according to the needs of each district and community Aligned with state, national, and international standards
- Can be accessed in community access points after-school and out-of-school

Problem: Lack of Culture

Despite the recognition that a change is needed in the way we educate children, shifting from rote learning and memorization to skill development that facilitates the application of knowledge and information, most classrooms have not changed. There is a need to shift the culture of schools, of students and families and communities to adapt and adopt an integrated STEM education. It is also critical that access to high-quality, STEM education is accessible to all students and not just some students. Currently, most school districts are lacking a comprehensive approach to implement an integrated, STEM-focused curriculum. In fact, one small change that could be made is the addition of computer programming courses in high school, yet only 51% of US high schools have computer [12] programming courses available to students. Schools in rural areas and schools with a greater percentage of minority or economically disadvantaged students being less likely to have access to [12] those courses. In districts with limited resources, access to STEM learning experiences may also be limited to students who are high-performing, or who have the resources to provide these experiences outside of school through after school programs. Access may also be limited due to lack of teacher preparedness to deliver STEM learning experiences, which limits the ability to develop a thriving STEM culture.

Studies point to the lack of educators' knowledge and confidence in delivering project-based [13,14] learning strategies. This should not come as a surprise as most educators have not been trained in these teaching strategies. Most STEM education studies have focused on specific subjects such as science and mathematics rather than looking at STEM as an integrated, cross [14, 15] discipline study and little understanding about the type of professional development is provided. It is clear that there are key components to professional development activities that promote[16-19] successful implementation in the classroom. The lack of a consistent definition of STEM, and vague studies of professional development practices has prevented the systematic studies needed [20] to identify teaching practices that significantly impact student outcomes. It has been shown that [21] teachers' beliefs, self-efficacy, and content knowledge do significantly affect student outcomes.

In addition to cultural shifts through professional development, educators must adjust their lesson plans and curricula in order to integrate STEM, PBL learning experiences. This shift takes a considerable amount of time for districts to adjust curricula and provide training to teaching staff to implement new curriculum in a classroom. This must be a focus of the district, and a shift in the culture of learning delivery in order to increase access to all students. Merely changing the curriculum and means of instruction is not sufficient to institute sustained system changes as it occurs in a near vacuum, without involvement from family members, friends and community members all of whom influence the student. To truly shift culture, we must do so in-school and out of-school, at home, and in the community.

Solution: Increase Design Thinking-Infused Pedagogy in Every Classroom

21stCentEd impacts the school district STEM teaching and learning culture by providing teacher LBDPD or Learn By Doing Professional Development for educators through design thinking infused pedagogy (DTIP). Students and teachers engage in co-learning STEM experiences that help students and teachers break down barriers and work together in ways that are dynamic and collaborative. Using a design thinking lens, educators learn how to develop a culture of collaborative inquiry, rigorous and relevant lessons/units/projects, academic literacy, and student agency.

The Implementation and Curriculum team at 21stCentEd is led by Dr. Jerome (JT) Taylor, who has designed a professional development training for educators to ensure they have the skills and tools necessary to successfully integrate STEM experiences into the classroom. Professional development is provided to administrators, educators, and individuals who can facilitate STEM™ experiences through the 21stCentEd Facilimentor trainings. The professional development follows a design thinking infused pedagogy (DTIP) as the framework, or context for educators. This professional development combined with the experiences in the classroom using the 21stCentEd [22] digital platform provides a solution to weaknesses that have existed in the past leaving teachers lacking access to experiences so students can practice and apply the design-thinking process. DTIP for teachers takes a deeper dive into applying design-thinking strategies to teaching and learning. Students benefit from a teaching and learning approach that considers their input, experience, interests, and creative problem-solving skills in real-world contexts. Lessons and activities afford flexible implementation methods allowing for course mentoring opportunities for teachers, students, parents, and community partners. This also allows for a flipped classroom model which, when paired with digital learning tools, has increased students' test scores and allows them to take on an active [23, 24] role in their learning. The team uses the four shifts protocol to help educators assess their current lesson plans and find opportunities to embed project-based and design-thinking experiences into their lessons alongside the 21stCentEd platform of STEM experiences.

Problem: Lack of Community

Community engagement can be conceptualized by considering the effects of the students' family as well as the broad community on the development of STEM skills, or STEM career aspirations. Prior research shows that students from marginalized communities find more value in STEM when they [25, 26] understand how STEM knowledge and skills can help solve issues affecting their communities. Finding more value in STEM may help minority students foster a sense of belonging in their present STEM classes and imagine themselves in STEM careers in the future. Additional studies show that [27] out-of-school activities, such as discussions of STEM subjects with caregivers and family members and watching shows that are STEM focused provide informal learning experiences that are [28] correlated with STEM career interests, even when formal STEM experiences are not. As such, it is critical to involve the families and community members to create a sustainable model that will encourage ongoing emphasis for STEM education that will create long-term benefits to the community.

School and community partnerships are critical to meet the ever-changing workforce demands. Community partnerships can facilitate learning, internships and real-world job experience that increase the likelihood of persistence in STEM career paths while filling critical workforce gaps. Additionally, community-based project work, and internships have demonstrated persistence in STEM degree attainment. Importantly, as technology continues to develop it is likely [29] that the skills needed in the workforce will shift rapidly, as previously mentioned. If community partners are engaged with the schools they can work together to ensure that curriculum continues to be aligned with the skills and talents needed in the workforce. The Collective Impact framework can be used to build the long-lasting, mutually trusting relationships that are necessary for cultural and social change [30].

Solution: Develop CISTEMIC - Collective Impact STEM Integrated Community Partnership

21stCentEd has also developed programming for community engagement through our CISTEMIC model. CISTEMIC stands for Collective Impact STEM Integrated Communities. Only through collective impact and community involvement can we ensure that we are educating youth with the skills and talents needed in the workforce that will ultimately enhance the economic vibrancy of the community. 21stCentEd has an experienced CISTEMIC team that works closely with district staff and community members. The work includes development of a district STEM strategy, stakeholder collaboration meetings, asset mapping, skills and talent needs analysis, STEM City proclamation, and an annual STEM Fest.

SHARED RESOURCES

STRATEGIC PLANNING

ASSET MAPPING

COMMUNITY PARTNERSHIPS AND STAKEHOLDER COLLABORATION

PROCLAMATION AS A STEM DISTRICT, STEM TOWN, OR STEM CITY

Additional Features

- District and community asset mapping to establish an inventory of current and needed resources in the development of a community-wide STEM initiative.
- Includes district-level and school-level capacity building, community organizing, and professional development services to assist districts in executing and transforming themselves into a STEM District, STEM-Town, or STEM-City by leveraging stakeholders in STEM talent development, workforce development, and economic development activities.
- District-led, community-wide 21st century STEM workforce and economic development initiatives to provide students real-world learning opportunities including localized internships, apprenticeships, and pathways to STEM-related fields.
- STEM Fest event bringing all community stakeholders together to engage in STEM activities and career opportunities. STEM-District, STEM-Town, or STEM-City proclamation event to announce the community's commitment to providing STEM learning opportunities for all students.

Impact Report

This impact report demonstrates 21stCentEd's commitment to its mission of "future-proofing ALL students for a world where STEM Literacy is at the center of our human existence." This commitment is centered on providing STEM experiences for all students, which is a robust, rigorous approach to STEM Education, where 21stCentEd leads the market.

Where is 21stCentEd Making an Impact?

The following data shows the number of students engaging in STEM experiences and the number of schools who have partnered with 21stCentEd to implement a Comprehensive STEM initiative. With the focus on STEM Education & learning recovery due to the pandemic, these numbers are rapidly growing. Importantly, 100% of students at partner schools have access to the 21stCentEd digital-learning platform.

Impact to Date

Eleven students who were enrolled in 21stCentEd Courses during the 2021- 2022 school year participated in a focus group at the end of the semester.

1/3 of the students completed an entrepreneurship course, 1/3 participated in a finance course, and 1/3 participated in an artificial intelligence course.

Main themes from focus group;

Students reported that the courses reflected their interests and were aligned with their expectations in terms of the content being what they needed/wanted.

- "The course met my expectations I learned a lot of important points on starting and running a business/product."
- "It honestly exceeded my expectations. I think it (entrepreneurship) opened my eyes to new things that I previously didn't think about when I initially thought about business related topics."

Students would recommend the courses to other students.

- "Yes because it gives insight on how to start, run, and succeed with a business/plan." "Yes because you can learn a lot of information from this and since it is not timed you can learn at your own pace."
- "I would recommend this course to other students who are also interested in STEM, coding, computer science, and AI because I have learnt a lot from the course and will be applying this knowledge later in life."

Students found the courses engaging.

- "the course is engaging because it has both video, reading, quiz, and projects... I found the videos more easier to pay attention to than the reading."
- "I found the course pretty engaging because there were many types of checks, quizzes, and reflections."
- "The course was engaging because it gave real life examples within the module. For example, it was easier to understand a topic when they created an exact scenario."

Students saw that the course relates to their future plans.

- "This course does relate to my future plans, because in more complex fields of physics, it is key to have artificial intelligence for telling patterns in results of experiments, because there are millions if not billions and trillions of results that must be categorized and analyzed for patterns and what researchers are looking for."
- "I wanna go into the medical field in the future and that requires going into medical school and that can be costly. Knowing how to better manage my money and how to save up could be useful to pay my student loans in the future."

Boxes We Check

Additional benefits of a Comprehensive STEM™ & CTE plan

✓ Distance/ Remote Learning

✓ Parent Involvement

✓ Blended Learning

✓ 21st-Century Skills

✓ Project-based Learning

✓ Social Emotional Learning

✓ STEM Teacher Shortage

✓ Equity

✓ Design/ Engineering Thinking

✓ Personalized Learning

✓ Learner-centered

✓ Learner Agency

✓ Digital Citizenship

✓ Empathy

✓ Student Engagement

✓ Career and College Readiness

✓ Peer mentoring

✓ Community Engagement

Key Performance Indicators:

21stCentEd is dedicated to future-proofing all students for a world where STEM Literacy is at the center of human existence. We do this through Comprehensive STEM™ and CTE experiences *EARLY*, *OFTEN*, and *EVERYWHERE*. The following key performance indicators help us measure this value proposition.

Early

- In the first year: At least 25-50% of students in an implementing school will have participated in 21stCentEd STEM experiences
- In the second year: At least 50-75% of students in the same implementing school will have participated in 21stCentEd STEM experiences.
- In the third year: At least 80% of students in the same implementing school will have participated in 21stCentEd STEM experiences.

Often

- By the end of the first year, 80% of participating students will have had multiple 21stCentEd experiences.
- By the end of the second year, at least 50% of students will participate in 21stCentEd STEM experiences multiple times a week.
- By the end of the third year, at least 80% of students will participate in 21stCentEd STEM experiences multiple times a week.

Everywhere

- By the end of the third year, at least 50% of participating students will have had 21stCentEd experiences in-school, after-school, or out-of-school
- By the end of the third year, student-serving community organizations will be set up as district partners and additional STEM access points for 21stCentEd experiences. The number and types of access points will vary due to location and circumstances.

Meaningful, Measurable Outcomes

21stCentEd has recently developed a robust evaluation plan to inform school districts about growth in STEM and 21st Century Skills as a result of implementing our comprehensive STEM and CTE programs. The LMS captures extensive data points including date, time, and length of time of activity for each student, grades, assignment completion, number of badges or assignments completed, number of courses completed, number of completion certificates earned. Additionally, each course will begin with a pre-completion assessment including outcome measures and end with the same assessment post-course completion.

Outcome Measures

Our LMS, Matrix, also allows for the embedding of surveys and assessments within specific courses for each student to complete as part of the outcome assessment. Upon initial entry into a course standardized, previously published and validated self-assessment measures will be completed by all students. These measures provide an initial assessment of the students' initial self-assessment of STEM and CTE engagement, identity, interest, knowledge, and problem solving. We use a pre-post methodology to compare responses prior to engagement, and at the completion of a course. Additionally, we use micro-surveys throughout the year to capture student's perceptions of the class as they progress through the course modules (sliding scale ranging from Like-Dislike) and their perception of the challenge level (sliding scale, Easy-Difficult). These data are helpful for our ongoing course improvement strategies.

We have deployed surveys containing items that have been demonstrated to align with each of our outcome measures. The survey contains 70 statements that are rated by students on a sliding scale with anchors strongly agree- strongly disagree (for 6th- 8th grade) and anchors (very true- not true) for elementary students. The survey takes approximately 15 minutes to complete. The assessments are also captured upon course completion. Additional outcomes can be captured through the LMS and through collaboration with school and/or district personnel to capture measures most critical for our partners.

"STEM Experience" Definition

Common STEM criteria across our courses that define a "STEM experience."

Design-thinking	Connected to careers in STEM-related fields
Project-based learning	Application of 21st century and STEM Literacy skills (entrepreneurship, the 4Cs, failing successfully, etc.)
Connected to real-world applications (not only theory)	Development and practice of empathy and other social emotional competencies
Interactive game-based and gamification-infused instruction	Adaptive Learning and Automation in an "Intelligent Learning System"

References

1. Li, Y., Wang, K., Xiao,Y., Froyd, J (2020). Research and trends in STEM education: a systematic review of journal publications. International Journal of STEM education. 7(11). doi: 10.1186/s40594-020-00207-6.

2. BLS. Occupational Outlook Handbook. Fastest Growing Occupations. bls.gov/ooh/fastest-growing.htm

3. The Condition of College and Career Readiness: National 2018," ACT, 2018

4. Mau, W. C. J., & Li, J. (2018). Factors influencing STEM career aspirations of underrepresented high school students. The Career Development Quarterly, 66(3), 246-258.

5. Pew Research Center (2018) Women and Men in STEM often at odds in the workplace. file:///Users/jodi-woodruff/Downloads/PS_2018.01.09_STEM_FINAL.pdf

6. US DOE. Final Priorities and Definitions-Secretary's Supplemental Priorities and Definitions for Discretionary Grants Programshttps://www.federalregister.gov/docu ments/2021/12/10/2021-26615/final-priorities-and-defini tions-secretarys-supplemental priorities-and-definitions-for

7. Deming, D., Noray, K. (2019) STEM Careers and the changing skill requirements of work. National Bureau of Economic Research. https://www.nber.org/papers/w25065

8. Harris, C. J., Penuel, W. R., D'Angelo, C. M., DeBarger, A. H., Gallagher, L. P., Kennedy, C. A., ... & Krajcik, J. S. (2015). Impact of project-based curriculum materials on student learning in science: Results of a randomized controlled trial. Journal of Research in Science Teaching, 52(10), 1362-1385.

9. Beier, M. E., Kim, M. H., Saterbak, A., Leautaud, V., Bish-noi, S., & Gilberto, J. M. (2019). The effect of authentic project-based learning on attitudes and career aspirations in STEM. Journal of Research in Science Teaching, 56(1), 3-23.

10. Dinh, T. V., & Zhang, Y. L. (2021). Engagement in high-impact practices and its influence on community college

transfers' STEM degree attainment. Community College Journal of Research and Practice, 45(11), 834-849.

11. Wang, L. H., Chen, B., Hwang, G. J., Guan, J. Q., & Wang, Y. Q. (2022). Effects of digital game-based STEM education on students' learning achievement: a meta-analysis. International Journal of STEM Education, 9(1), 1-13.

12. National Center for Education Statistics. Retrieved from:Code.org on 9.1.2022. https://docs.google.com/docu ment/d/1gySkItxiJn_vwb8HIIKNXqen184mRtz DX12cux0ZgZk/pub#h.oubb4frxpvkd

13.Shernoff, D., Sinha, S., Bressler, D., Ginsgburg, L. (2017) Assessing teacher education and professional development needs for the implementation of integrated approaches to STEM education. Journal of STEM education. 4(13). DOI: 10.1186/s40594-017-0068-1

14. Kelley, T.R., Knowles, J.G. (2016) A conceptual framework for integrated STEM education. IJ STEM Ed 3, 11. https://doi.org/10.1186/s40594-016-0046-z

15. Takeuchi,M., Sengupta,P., Shanahan,MC., Adams, JD, & Hachem,M. (2020) Transdisciplinarity in STEM education: a critical review, Studies in Science Education, 56:2, 213-253, DOI: 10.1080/03057267.2020.1755802

16. Darling-Hammond, L., Wei, R., Andree, A., Richardson, N., Orphanos, S. (2009). Professional Learning in the Learning Profession: A Status Report on Teacher Development in the United States and Abroad. National Staff Development Council.

17. Desimone, L. (2009). Improving impact studies of teachers' professional development: Toward better conceptualizations and measures. Educational Researcher, 38, 181-199.

18. Rhoton J. & Wohnowski, B. (2005) Building on going and sustained professional development. In J. Rhoton, and P. Shane (Eds.), Teaching Science in the 21st Century. National Science Teachers Association and National Science Education Leadership Association: NSTA Press.

19. National Comprehensive Center for Teacher Quality. (2011). Recruiting staff and attracting high-quality staff to hard-to-staff schools. In C. L. Perlman & S. Redding (Eds.), Handbook on Effective Implementation of School Improvement Grants. Charlotte, NC: Information Age.

20. Roehrig, G. H., Dare, E. A., Ellis, J. A., & Ring-Whalen, E. (2021). Beyond the basics: a detailed conceptual framework of integrated STEM. Disciplinary and Interdisciplinary Science Education Research, 3(1), 1-18.

21. Carney, M. B., Brendefur, J. L., Thiede, K., Hughes, G., & Sutton, J. (2016). Statewide mathematics professional development: Teacher knowledge, self-efficacy, and beliefs. Educational Policy, 30(4), 539-572.

22. Lin, K. Y., Wu, Y. T., Hsu, Y. T., & Williams, P. J. (2021). Effects of infusing the engineering design process into STEM project-based learning to develop preservice technology teachers' engineering design thinking. International Journal of STEM Education, 8(1), 1-15.

23. Campillo-Ferrer, J. M., & Miralles-Martínez, P. (2021). Effectiveness of the flipped classroom model on students' self-reported motivation and learning during the COVID-19 pandemic. Humanities and Social Sciences Communications, 8(1), 1-9.

24. McLeod, S., Graber, J. (2018).*Harnessing Technology for Deeper Learning (A Quick Guide to Educational Technology Integration*

and Digital Learning Spaces). Amazon. ISBN-13: 978-1943874088.

25. McGee, E., & Bentley, L. (2017). The equity ethic: Black and Latinx college students reengineering their STEM careers toward justice. American Journal of Education, 124(1), 1-36.

26. Gray, D. L., McElveen, T. L., Green, B. P., & Bryant, L. H. (2020). Engaging Black and Latinx students through communal learning opportunities: A relevance intervention for middle schoolers in STEM elective classrooms. Contemporary Educational Psychology, 60, 101833.

27. Kang, H., Calabrese Barton, A., Tan, E., D Simpkins, S., Rhee, H. Y., & Turner, C. (2019). How do middle school girls of color develop STEM identities? Middle school girls' participation in science activities and identification with STEM careers. Science Education, 103(2), 418-439.

28. Edmonds, J. Lewis, F., & Fogg-Rogers, L. (2022) Primary pathways: elementary pupils' aspiration to be engineers and STEM subject interest, International Journal of Science Education, Part B, DOI: 10.1080/21548455.2022.2067906

29. Wackler, T. Et al. (2018) Charting a course for success: America's Strategy for STEM education. Committee on STEM education. National Science and Technology Council. https://www.energy.gov/sites/default/files/2019/05/f62/STEM-Education-Strategic-Plan-2018.pdf

30. Brian D. Christens & Paula Tran Inzeo (2015) Widening the view: situating collective impact among frameworks for community-led change, Community Development, 46:4, 420-435, DOI: 10.1080/15575330.2015.1061680

About the Publisher

CISTEMIC Publishing is an imprint of **21stCentEd**, a
technology company committed to providing a
ComprehensiveSTEM^{TM} education to students early, often,
and everywhere.
Please visit https://cistemic.us/ to learn more.

Read More from The STEM Century Series

STEM Century: It Takes a Village to Raise a 21st Century Graduate

STEM Century: The National Alliance of Black School Educators Edition